Experience the California Coast
Beaches and Parks from Monterey to Ventura

COUNTIES INCLUDED

MONTEREY · SAN LUIS OBISPO · SANTA BARBARA · VENTURA

Experience the California Coast
Beaches and Parks from Monterey to Ventura

COUNTIES INCLUDED
MONTEREY · SAN LUIS OBISPO · SANTA BARBARA · VENTURA

State of California
Arnold Schwarzenegger, *Governor*

California Coastal Commission

Peter Douglas, *Executive Director*   Susan Hansch, *Chief Deputy Director*

Steve Scholl
Editor and Principal Writer

Erin Caughman
Designer and Co-Editor

Jonathan Van Coops
Mapping and GIS Program Manager

Gregory M. Benoit
Principal Cartographer

Caitlin Bean   Mark Johnsson, Ph.D.
John Dixon, Ph.D.   Sylvie B. Lee
Lesley Ewing, P.E.   Audrey McCombs
Lillian Ford   Chris Parry
Ben Hansch   Jonathan Van Coops
Contributing Writers

Jo Ginsberg
Consulting Editor

Artwork by Tom Killion

Linda Locklin
Coastal Access Program Manager

University of California Press
Berkeley   Los Angeles   London

University of California Press
Berkeley and Los Angeles, California

University of California Press, Ltd.
London, England

Library of Congress Cataloging-in-Publication Data

Beaches and parks from Monterey to Ventura : counties included,
Monterey, San Luis Obispo, Santa Barbara, Ventura / Steve Scholl,
editor and principal writer ... [et al.].
    p. cm. -- (Experience the California coast guides)
  "California Coastal Commission."
  Includes bibliographical references and index.
  ISBN-13: 978-0-520-24949-3 (paper : alk. paper)
  1. Pacific Coast (Calif.)--Guidebooks. 2.
California--Guidebooks. 3. Beaches--California--Pacific
Coast--Guidebooks. 4. Parks--California--Pacific Coast--Guidebooks.
I. Scholl, Steve. II. California Coastal Commission.
  F868.P33B43 2007
  917.9404'54--dc22

                              2006029129

Printed in Italy by Grafiche SIZ S.p.A.

15 14 13 12 11 10 09 08 07

10 9 8 7 6 5 4 3 2 1

Contents

Santa Barbara County
181–252

Ventura County
253–299

Features

Big Sur coast

Introduction

ON CALIFORNIA'S Central Coast you will find some of the world's most spectacular coastal sights. Beaches and coastal parks in Monterey, San Luis Obispo, Santa Barbara, and Ventura Counties include many of the state's most popular, with facilities ranging from the plain and simple to Hearst Castle, a mansion of unparalleled excess. The Central Coast's lesser known beaches and recreation areas are also described in this guide, offering to every visitor a favorite spot on the shore.

This book is intended to depict the richness and diversity of the California coast. It is meant for all coastal visitors, whether equipped with beach blanket, binoculars, or bodyboard. The book tells you where to find over 300 beaches, parks, and other recreational facilities along the Central Coast, what coastal resources are at each location, and what you might do there. Most of the sites listed here are publicly owned, while others are privately managed.

Point Conception, located in Santa Barbara County, is the intersection of northern and southern California. North of Point Conception, powerful breakers pound the rocky shore, and tidepools invite exploration. Beaches of crystalline whiteness, and of dark-hued pebbles strewn with moonstones, are scattered along the coast. The pine and cypress trees of Monterey and San Luis Obispo Counties have come to evoke for many visitors the shoreline of this part of California.

South of Point Conception are chaparral-covered slopes and warm, sandy beaches. Sheltered ocean waters and the rays of the sun bring to mind sailboats and sand castles. World-famous surf breaks punctuate the shore. Palm trees, although non-native, make themselves right at home. Water sports are highly popular at Santa Barbara and Ventura County beaches, where average ocean water temperatures in summer approach 70 degrees Fahrenheit.

All along the four counties described in this book, air temperatures are generally mild. In the north, average high temperatures range from near 60 degrees Fahrenheit in winter to the low 70s in summer; in the south, average highs range from the mid 60s in winter to the high 70s in summer. Coastal overcast sometimes moderates high temperatures in spring or summer; the autumn months are often the warmest. And although the seasons are indeed distinguishable on this Mediterranean-type coast, with its rainless summers and cool, sometimes wet winters, a warm sun and cloudless sky could frame your coastal visit on just about any date in the calendar.

The California Coastal Commission, along with the State Coastal Conservancy, the Department of Parks and Recreation, and the Department of Fish and Game, is charged with conserving, enhancing, and making available to the public the beaches, accessways, and resources of the coast. The Coastal Commission's responsibilities under the law known as the California Coastal Act include providing the public with a guide to coastal resources and maintaining an inventory of paths, trails, and other shoreline accessways available to the public. This book furthers those purposes, as does the first book in the California Coastal Commission's new guidebook series, *Experience the California Coast: A Guide to Beaches and Parks in Northern California*, and the previously published *California Coastal Resource Guide* and *California Coastal Access Guide*.

This guide does not encourage trespass on private property. Not included here are a few commonly used trails over which the public lacks legal right of access. Informal beach paths that cross the right-of-way of the Union Pacific Railroad, as in a few locations in Santa Barbara County, may involve trespass if they lack public use easements. The California Coastal Commission, the State Coastal Conservancy, and local governments in the four-county region addressed by this book continue to press for increased opportunities for legal, safe access to the beach. We have attempted to include all known public beaches and accessways; as additional shoreline accessways become available, they will be included in future editions of the California Coastal Commission's guides.

This book lists, for smaller communities, commercial outfitters that sell or rent kayaks, surfing equipment, bicycles, and other gear. In the more populous areas, these facilities are too numerous to include in this book; check local yellow pages or Internet search services to find what you need. The information here is as complete as space allows; call ahead to make sure the recreational offerings you seek are available. The editors welcome suggestions for future editions (see p. 300).

For an economical overnight stay, this guide lists hostels, state and local campgrounds, and, as space permits, private campground facilities. Campsites in public or private parks include family camps, group camps, sites with RV hookups, walk-in environmental campsites, hike or bike sites, and enroute (overflow) spaces. Many can be reserved in advance. Where private campgrounds are too numerous to be listed individually, visitors are directed to clearinghouses such as the local chambers of commerce; see the introduction for each county.

In addition to having one of the most striking shorelines found anywhere in the world, the coast of Monterey, San Luis Obispo, Santa Barbara, and Ventura Counties holds distinctive coastal villages and urban attractions that equal those of metropolitan areas far larger in size. Information about market-rate hotels, inns, eating establishments, and other visitor destinations is available in numerous other guidebooks.

Enjoy your visits to California's spectacular coast. Keep safe by observing posted restrictions along hazardous stretches of shoreline. Remember that sleeper waves are a factor on the California coast. When strolling the beach or checking out tidepools, make it a general rule not to turn your back on the ocean. Remember that large waves may wash over what look like safe spots on rocks and bluffs.

Natural conditions along the California coast are always changing, and the width of beaches and shape of bluffs can be altered by the seasonal movement of sand or by erosion. Coastal access and recreation facilities can be damaged by these forces, and trails, stairways, parking areas, and other facilities may be closed for repairs. When planning any trip to the coast, check ahead of time to make sure that your destination is currently accessible. Some facilities, such as park visitor centers, are run by volun-

For general information on state parks, including a list of camping and day-use fees and campgrounds available without a reservation, see www.parks.ca.gov.

For state park camping reservations, call: 1-800-445-7275 (available 24 hours), or see www.reserveamerica.com.

For other camping opportunities, see individual entries that follow.

For information on Hostelling International's facilities, see www.hiayh.org.

teers and are open only limited hours; call ahead to check open times. Facilities such as running water are limited or not available at some parks and shoreline accessways; it is a good idea on a coastal trip to bring water, food, waterless hand cleaner, and an extra layer of clothing. Key information is included here about public transit lines that serve beaches in the larger communities described in this book; check with local transit providers for details.

Dogs also enjoy coastal outings, but their inquisitive nature can create hazards for coastal wildlife. In state parks, dogs must be kept on leashes that are no more than six feet long and in a tent or enclosed vehicle at night. Except for guide dogs, pets are not allowed in state park buildings, on trails, or on most beaches. Although allowed in some city and county beach parks, dogs may be subject to leash requirements. See individual site descriptions, and please observe posted signs regarding dogs on trails and beaches and in parks.

This guide's purpose is to contribute to a better understanding of the importance of coastal resources, both to the quality of life for people and to the maintenance of a healthy and productive natural environment. This book is offered with the knowledge that a wide appreciation for the coast among Californians plays an important role in the protection and restoration of coastal resources.

Playing in the surf

Using This Guide

Each group of sites is accompanied by a map and a chart of key facilities and characteristics. The "Facilities for Disabled" chart category includes wheelchair-accessible restrooms, trails, campsites, or visitor centers; text descriptions note where restrooms are not wheelchair accessible. The "Fee" category refers to a charge for entry, parking, or camping. Check the index for surfing spots, beaches with lifeguard service, and other recreational highlights. Most parks and recreational outfitters maintain websites, but URL addresses may change and space in the book is limited; use any popular Internet search engine to look for more information on facilities listed in this guide.

Brief introductions to coastal environments such as dunes, rocky shore, and Monterey pine forest are included, along with highlights of plants, animals, and other resources that you may see there. For more information about the California coast, consult the Bibliography and Suggestions for Further Reading (p. 308).

Sandy Beach / Rocky Shore / Trail / Visitor Center / Campground / Wildlife Viewing / Fishing or Boating / Facilities for Disabled / Food and Drink / Restrooms / Parking / Fee

Sand play

Caring for the Coast

THIS BOOK helps you get the most out of visiting California's Central Coast, which offers endless enjoyment, beauty, solace, and adventure. But what have you done for the coast lately? You can contribute to its good health by developing an awareness of how it is affected by your everyday actions, and by striving to act in ways that will have beneficial results. Here are some tips. For more ideas and to take the Coastal Stewardship Pledge, visit www.coastforyou.org or call 1-800-COAST-4U.

Stash Your Trash

Researchers have found alarming quantities of plastic debris in the open ocean, where it circulates continuously unless and until it is consumed by a bird, fish, or marine mammal. Most of this debris comes from land, and was carried to the ocean by rain, tides, or wind. Avoid contributing to this problem by always disposing of trash properly, and by practicing the three "Rs"—reduce the waste you generate, buy reusable items, and recycle trash when possible. When going to the beach or out on a boat, bring a bag and pick up the debris you come across. Each piece you collect is one less hazard for a marine animal. Another way to help is to volunteer for a beach cleanup activity such as Coastal Cleanup Day or the Adopt-A-Beach Program.

Watch Your Step!

Certain types of coastal habitats and the wildlife that live there are especially sensitive to human encounters. To minimize your disturbance to these ecologically important places, please observe the guidelines that follow.

Volunteers participating in a beach cleanup

Tidepool observation

Coastal Dunes and Wetlands

Dunes and wetlands are susceptible to damage if vegetation is disturbed by foot traffic or off-road vehicles. Sensitive wildlife species are vulnerable to disturbance by humans, dogs, and horses. When visiting dunes and wetlands, stay on prescribed pathways or boardwalks and pay close attention to rules imposed by land managers. Where dogs are allowed, keep them leashed.

Snowy Plovers

The western snowy plover is a threatened species that nests on certain sandy beaches, including those backed by dunes. Help protect this small bird by steering clear of nests and generally keeping to the wet sand. Avoid fires on the beach, leave driftwood and kelp in place, and dismantle driftwood structures that provide perches for birds that prey on plovers and their chicks. Keep dogs leashed where allowed. Several Central Coast beaches, such as the Coal Oil Point Reserve in Santa Barbara County, have docents who inform the public about snowy plovers – look for them (or join them).

Tidepool Etiquette

Tidepool plants and animals are fascinating to view, but they can be harmed by human contact. When visiting tidepools:

- Watch where you step. Step only on bare rock or sand.

- Don't touch any living organisms. A coating of slime protects most tidepool animals, and touching them can damage them.

- Don't prod or poke tidepool animals with a stick. Don't attempt to pry animals off of rocks.

- Leave everything as you found it. Collecting tidepool organisms is illegal in most locations and will kill them. Cutting eelgrass, surfgrass, and sea palm is prohibited.

Watching Wildlife

Observing wild animals in their natural environment is a rare treat. To ensure that the encounter results in no harm to either the animal or the human observer, follow these guidelines:

* Keep your distance. Maintain enough distance so the animal is not aware of you. One hundred yards is a good rule of thumb, although some animals require more (500 yards is the guideline for viewing whales from a boat). If your presence causes an animal to change its behavior—even if it just looks at you—move away immediately. Use binoculars or zoom lenses for a close-up look.

* Watch quietly and limit the time spent observing animals. Encounters with people can be stressful to animals. Half an hour is reasonable.

* Stay clear of mothers with young. Nests, dens, and rookeries are especially vulnerable to human disturbance.

* Resist the temptation to "save" animals. If an animal appears sick, get help from a professional. In Monterey and San Luis Obispo Counties, call 415-289-SEAL; in Santa Barbara County, call 805-687-3255; in Ventura County, call 805-672-4947.

* Never surround an animal. Avoid approaching wildlife directly and always leave an escape route. Never trap an animal between a vessel and shore.

* Leave pets at home, or keep them on a leash and away from wildlife.

* Never feed wild animals.

* Report illegal poaching (or polluting) to the authorities: call 1-888-DFG-CALTIP. *The program is confidential and you may be eligible for a reward.*

Sensible Seafood Choices

Use your purchasing power to support healthy oceans by selecting seafood that is harvested in a sustainable and environmentally responsible manner. For a pocket guide to sensible seafood choices, visit: www.montereybayaquarium.org/cr/seafoodwatch.asp.

Non-point Source Pollution

Another way that people affect the health of the coast is through non-point source pollution, which gets flushed into the ocean by stormwater runoff. Minimize your contribution to this problem by taking simple actions; for example, use least-toxic gardening products, maintain your car to prevent oil leaks, and pick up after your dog.

Whale Tail License Plate

California drivers can help the coast by purchasing a Whale Tail License Plate. The plate funds coastal access trails, beach cleanups, and marine education throughout California, including grants to local groups. Call: 1-800-COAST-4U, or visit www.ecoplates.com.

Map Legend

TRANSPORTATION

———————————— Major Road
———————————— Minor Road
————●1●———— California State Highway
————🛡101🛡———— United States Highway
————●580●———— Interstate
·—·—·—·—·—·—·—· Railroad

SHORELINE AND HYDROGRAPHY

———————— Shoreline
———————— Rivers and Streams

Pacific Ocean, Bays Lakes, and Ponds

TRAILS AND BIKE WAYS

Hiking Trail
Hiking Trail Along State Highway
Hiking Trail Along Road

Pacific Coast Bicentennial Bike Route
Bike Route Along State Highway
Bike Route Along Road

TOPOGRAPHY AND BATHYMETRY

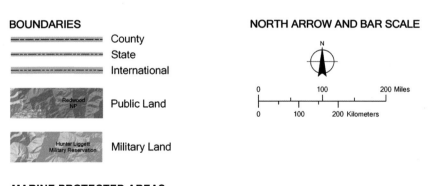

0   500 1000 2000 4000 8000 11000 14500 Feet
0   150 300 600 1200 2400 3350 4400 Meters
Elevations approximate

Bathymetry
200 meter interval

-16000   -12000   -8000   -4000   0 Feet
-4800   -3600   -2400   -1200   0 Meters
Depths approximate

BOUNDARIES

══════════════ County
══ ═ ══ ═ ══ State
═══════════════ International

Redwood NP — Public Land

Hunter Liggett Military Reservation — Military Land

NORTH ARROW AND BAR SCALE

N

0   100   200 Miles
0   100   200 Kilometers

MARINE PROTECTED AREAS

At publication, the status of Marine Protected Areas located offshore along the Central Coast is under review by the California Department of Fish and Game. For information, contact the Department's Marine Region in Monterey at 831-649-2870.

DATA AND INFORMATION SOURCES

California Coastal Commission
California Department of Fish and Game
California Department of Parks and Recreation
California Spatial Information Library

Greeninfo Network
Monterey County
Santa Barbara County
U.S. Geological Survey

Page opposite: Northern Big Sur, Monterey County

Monterey County

Monterey County

THE COAST of Monterey County occupies a prominent place in California's history and among the state's natural and scenic marvels. The Ohlone, or Costanoan, Native Americans occupied the Monterey Bay area for at least 10,000 years prior to Spanish settlement. The Big Sur Coast was occupied by the Ohlone, Salinan, and Esselen Indians, the latter one of California's smaller indigenous societies and the first to become extinct. Spanish explorer Juan Rodríguez Cabrillo sighted the Monterey Peninsula in 1542, and Monterey Bay was apparently entered by Sebastián Rodríguez Cermeño, a Portuguese explorer who made his way south along the coast in a small boat in December 1595, after his galleon *San Agustín* was wrecked on Point Reyes. Sebastián Vizcaíno, who had accompanied Cermeño, returned in 1602 to make a landing on the shores of Monterey Bay, a harbor that he described as so magnificently sheltered that later Spanish explorers had trouble recognizing it. In 1769, Juan Gaspar de Portolá's pioneering overland expedition set out from San Diego to find Monterey Bay, but went past it, encountering San Francisco Bay before heading back south. The following year, Portolá and Padre Junípero Serra returned to establish the Presidio of Monterey and the second California mission, San Carlos Borroméo, at Carmel.

The county's coast offers some of the world's best-known coastal attractions. Beaches include the unbroken stretch of sand along the great arc of Monterey Bay, the brilliant white sand of Carmel, and the scattered beaches of Big Sur, some made inaccessible by the sheer cliffs of the Coast Range. Monterey pine and cypress trees, sculptured by wind in their dramatic settings, grow naturally on a tiny band of California coast centered on the Monterey Peninsula, but are now planted widely around the globe. The sea otters of the Monterey County coast are winsome, but feisty, marine mammals.

Coastal communities in Monterey County vary dramatically, reflecting their settings. Monterey overlooks its sheltered bay, while Carmel and Pacific Grove are towns built in the forest, and Marina is a town in the dunes. Big Sur is not really a town at all, but a thinly settled patch of mountain and shore with a reputation built on its mix of dramatic scenery and personal exploration.

Monterey-Salinas Transit serves Monterey-area communities and Hwy. One to Big Sur; the MST Trolley links downtown Monterey, Pacific Grove, and the Monterey Bay Aquarium. The Monterey Transit Plaza is at Munras Ave. and Alvarado St.; call: 831-899-2555.

Monterey Visitor Center, Camino El Estero and E. Franklin St., 831-649-1770.

Monterey County Convention and Visitors Bureau, Olivier St. near Scott St., 831-657-6400.

Moss Landing Chamber of Commerce, 831-633-4501.

Marina Chamber of Commerce, 831-384-9155.

Seaside/Sand City Chamber of Commerce, 831-394-6501.

Pacific Grove Chamber of Commerce, 831-373-3304.

Carmel Visitor Center, 831-624-2522.

Big Sur Chamber of Commerce, 831-667-2100.

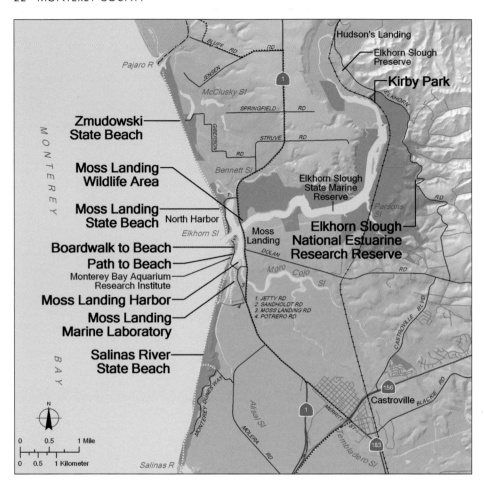

Hudson's Landing

Elkhorn Slough
Preserve

Kirby Park

Pajaro R

BLUFF RD
RD
JENSEN
McClusky Sl

SPRINGFIELD RD

GIBERSON
STRUVE RD

RD

ELKHORN

**Zmudowski
State Beach**

**Moss Landing
Wildlife Area**

Bennett Sl

Elkhorn Slough
State Marine
Reserve

Parsons
Sl

RD

**Moss Landing
State Beach** North Harbor

Elkhorn Sl

1

Boardwalk to Beach

Moss
Landing

**Elkhorn Slough
National Estuarine
Research Reserve**

Path to Beach

DOLAN

RD

Monterey Bay Aquarium
Research Institute

Moro Cojo

Sl

Moss Landing Harbor

**Moss Landing
Marine Laboratory**

1. JETTY RD
2. SANDHOLDT RD
3. MOSS LANDING RD
4. POTRERO RD

CASTROVILLE BLVD

**Salinas River
State Beach**

M O N T E R E Y

B A Y

N

0   0.5   1 Mile
0   0.5   1 Kilometer

MONTEREY DUNES WAY

Alisal Sl

MOLERA RD

156

1

Castroville

BLACKIE RD

MERRITT ST

Embladero Sl

183

Salinas R

Moss Landing Marine Laboratory

Northern Monterey County

	Sandy Beach	Rocky Shore	Trail	Visitor Center	Campground	Wildlife Viewing	Fishing or Boating	Facilities for Disabled	Food and Drink	Restrooms	Parking	Fee
Zmudowski State Beach	•					•	•			•	•	
Kirby Park			•			•	•	•		•	•	
Elkhorn Slough National Estuarine Research Reserve			•	•		•		•		•	•	•
Moss Landing Wildlife Area			•			•				•		
Moss Landing State Beach	•					•	•			•	•	
Boardwalk to Beach	•							•		•		
Path to Beach	•											
Moss Landing Harbor					•	•	•	•	•	•	•	
Moss Landing Marine Laboratory				•		•		•		•	•	
Salinas River State Beach	•		•			•	•			•	•	

ZMUDOWSKI STATE BEACH: *End of Giberson Rd., 3 mi. N. of Moss Landing.* From Hwy. One, turn west on Struve Rd., then continue on Giberson Rd. two miles through farm fields. Zmudowski State Beach offers an undeveloped expanse of dunes and a beach popular for surf fishing and beachcombing. A slough to the north of the parking area can be viewed from an adjacent boardwalk. On the beach, marbled godwits feed along the surf line, and brown pelicans fly low over the ocean.

Zmudowski State Beach is part of the miles-long crescent, backed by dunes, that forms the eastern shore of Monterey Bay. It is possible to walk or ride a horse along the wet sand all the way to the harbor entrance at Moss Landing. The parking area is open from 8 AM until a half hour after sunset. Portable toilets. Day use only; no dogs or fires allowed. Horses are permitted on the wet sand only. Call: 831-649-2836.

KIRBY PARK: *Elkhorn Rd., N.E. of Moss Landing.* A public fishing access is located at the northeast end of the slough. There is a ramp and floating pier; this is one of the few places (Moss Landing Harbor is another) to launch a boat or kayak into Elkhorn Slough. Portable restrooms and barrier-free access to the waterfront for fishing. A mile-long paved wheelchair-accessible trail runs north along the slough. Open from 5 AM to 10 PM daily. Managed by the Moss Landing Harbor District; call: 831-633-2461.

ELKHORN SLOUGH NATIONAL ESTUARINE RESEARCH RESERVE: *1700 Elkhorn Rd., N. of Castroville.* The 1,400-acre reserve protects wetlands and uplands for scientific research, public education, and visitor enjoyment. Trails lead through stands of coast live oak trees and grasslands to viewpoints overlooking wetland fingers and the main channel of Elkhorn Slough. At the Parson's Slough overlook, 116 bird species were seen on a single day in October 1982, setting a North American record. The visitor center near the main entrance has interpretive displays, a bookstore, picnic tables, and restrooms. Binoculars and bird books are available for loan.

A wheelchair-accessible trail leads one-third mile from the visitor center to an overlook with a broad view over the Elkhorn Slough complex. Docents lead nature walks on weekends. Entrance to the visitor center and picnic area are free. Fee charged for entry into the reserve; free entrance with valid fishing or hunting license. The reserve is open from 9 AM to 5 PM, Wednesday through Sunday. No pets allowed; fires, camping, boating, and firearms are prohibited. The Elkhorn Slough National Estuarine Research

Moss Landing Harbor

Reserve is managed cooperatively by the California Department of Fish and Game in partnership with the National Oceanic and Atmospheric Administration, with support from the nonprofit Elkhorn Slough Foundation. The waters of the slough are part of the Monterey Bay National Marine Sanctuary. Call: 831-728-2822.

MOSS LANDING WILDLIFE AREA: *N. of Elkhorn Slough, E. of Hwy. One, Moss Landing.* Watch carefully for the abrupt turn-off from Hwy. One, marked by a brown sign with a binocular icon that indicates wildlife viewing. Park to the east of the boat repair shop. The wildlife area protects nesting sites for the western snowy plover and roosting grounds for other birds. A trail leads from the small parking area to an overlook on the north side of Elkhorn Slough; in summer, watch for brown pelicans cruising low over the water.

MOSS LANDING STATE BEACH: *Jetty Rd., W. of Hwy. One.* Fishing, surfing, strolling, and horseback riding are popular on this narrow spit between the ocean and Bennett Slough. Along Jetty Rd., there are three pull-outs, each with limited parking, restrooms, and a short path to the beach. Yellow and pink sand verbena can be seen growing in the dunes. Strong rip currents make the ocean unsafe for swimming. For picnickers, the dunes provide some shelter from afternoon winds.

The area is a good place to view birds, including surf scoters, brown pelicans, sanderlings, killdeers, and marbled godwits. On exposed mudflats to the east of Jetty Rd., shorebirds, gulls, and terns congregate. A large parking area is located at the end of Jetty Rd., overlooking the harbor. Kayaks and canoes can be launched here into Elkhorn Slough. Look for sea otters feeding in the narrow entrance channel or pigeon guillemots resting on the water; harbor seals haul out on the sand. Moss Landing State Beach offers day use only; no fires.

BOARDWALK TO BEACH: *7544 Sandholdt Rd., Moss Landing.* A wheelchair-accessible boardwalk to the sandy beach is located on the north side of a Moss Landing Marine Laboratory building. Parking; no facilities.

PATH TO BEACH: *N. of 7700 Sandholdt Rd., Moss Landing.* A half-hidden path located between a marine laboratory building and Phil's Fish Market leads 200 feet to the beach. Loose sand path; no facilities.

MOSS LANDING HARBOR: *Sandholdt Rd., W. of Hwy. One, Moss Landing.* The harbor is used for commercial and sport fishing, research vessels, and pleasure boating, and is considered to be an extremely safe refuge. Facilities include slips and dry storage, fuel dock, pumpout station, supplies, bait and tackle, restaurants, showers, and restrooms. The mouth of Elkhorn Slough separates the harbor into two parts. The North Harbor has a two-lane boat ramp, just west of Hwy. One; fee charged. The South Harbor includes parking and fishing and is the site of the harbor office. Call: 831-633-2461. The privately operated Moss Landing Recreational Vehicle Park, adjacent to the South Harbor parking area, offers 46 RV spaces with full hookups; call: 877-735-7275. The Monterey Bay Aquarium Research Institute is located in the harbor area; no public facilities.

MOSS LANDING MARINE LABORATORY: *8272 Moss Landing Rd., Moss Landing.* This facility serves marine researchers and students, but the lobby houses a small visitor center. Tours of the laboratory are available for groups of ten or more. Field trips for school groups and occasional programs for the public are also offered. In front of the laboratory, a native plant area is landscaped with sticky monkeyflower and California sagebrush. A wheelchair-accessible boardwalk offers a pleasant view of the slough; telescopes are available to look for harbor seals, egrets, and cormorants. Open 8 AM to 5 PM, Monday through Friday; for information, call: 831-771-4400.

SALINAS RIVER STATE BEACH: *W. of Hwy. One, Moss Landing.* This state park includes two miles of sandy beach, backed by vegetated dunes. There are three separate parking areas and beach access points. The northernmost parking area is located at the south end of Sandholdt Rd., near Moss Landing Harbor. A short, loose-sand trail leads through vegetated dunes to a steep beach, which is part of the seemingly endless curve of Monterey Bay shoreline. A second parking area and path are located at the end of Potrero Rd. These two parking lots are linked by a trail, overlooking the slough that was formerly a part of the Salinas River, before that stream carved a new exit farther south. Watch for brush rabbits moving warily through the dunes, or white-crowned sparrows feeding in groups.

A third beach access point in Salinas River State Beach is reached via Molera Rd. and Monterey Dunes Way. A loose-sand, quarter-mile-long path, longer than the others at this state park unit, leads through vegetated dunes to a steep beach that is popular for fishing, clamming, and hiking. The large paved parking area is suitable for horse trailers, and equestrians use this as a starting point for riding along the shore; riders are limited to the beach and the trail through the dunes. Fires not allowed. For information, call: 831-649-2836.

Salinas River State Beach

Maritime Chaparral

MARITIME chaparral, a unique shrubby plant community with a high degree of endemism, occurs in patches along the coast of California from Sonoma County to San Diego County. It is composed of low-growing woody shrubs and is generally restricted to well-drained, coarse-textured sandy soils within the fog belt. Some of the common shrub species that occur in maritime chaparral are manzanitas (*Arctostaphylos* sp.), live oaks (*Quercus* sp.), chamise (*Adenostema fasciculatum*), toyon (*Heteromeles arbutifolia*), wild lilac (*Ceanothus* sp.), coffeeberry (*Rhamnus* sp.), and coast silk tassel (*Garrya elliptica*). All of these shrubs are evergreen, and they generally have small, leathery leaves with a thick cuticle, which is the wax-like layer covering the leaf. Adaptations such as these help these species thrive in the Mediterranean climate of coastal California.

Maritime chaparral is different from interior chaparral by having greater exposure to summer fog, humidity, and mild temperatures. The fog moderates the effects of summer drought. Even though both chaparral communities are fire adapted, maritime chaparral may not require fire at the same frequency as inland chaparral. However, maritime chaparral definitely needs fire. In its absence, maritime chaparral shrubs form a nearly impenetrable cover, and the community's diversity is compromised. Indeed, many plant species in maritime chaparral depend upon fire for their regeneration. Research has shown that fire in maritime chaparral greatly reduces the cover of coast live oak and manzanitas, and provides an opportunity for a host of other species to thrive, thus increasing biodiversity. This is a dynamic plant community; after a fire, a whole suite of herbaceous plants may be found growing in the new openings between the shrubs.

The California Native Plant Society has identified over a dozen plant species in this habitat as rare and many of them are given protection under the federal and state Endangered Species Acts. Before urban and agricultural development along California's coast displaced much of it, maritime chaparral dominated sandy landscapes near Monterey Bay, the Santa Cruz Mountains, Nipomo Mesa in San Luis Obispo County, and Burton Mesa in Santa Barbara County, among other places. Much of the original extent of this plant community has been eliminated.

Maritime chaparral

Chamise (*Adenostema fasciculatum*) is a perennial shrub common on dry slopes in the maritime chaparral and chaparral communities. The reddish to gray-brown bark begins peeling with age. Between May and June, white flowers bloom in sprays at the ends of the branches. The leaves are thickly clustered and quite resinous. Another common name for chamise is greasewood, because its branches and leaves contain highly flammable resins that catch fire easily, burn intensely, and contribute to the spread of brush fires. Chamise has a well-developed basal burl that quickly resprouts after a fire and holds the soil on hillsides. Chamise becomes dormant during the hottest, driest period of summer, and sheds both branches and bark in an effort to reduce the amount of tissue requiring moisture. These sloughed materials then serve as fuel for the next fire. Chamise is thought by some to be allelopathic, meaning that it reduces competition from other plants by releasing toxins into the soil that inhibit or prevent their growth.

Chamise

Woollyleaf manzanita (*Arctostaphylos tomentosa*) is a three-to-five-foot-tall evergreen shrub in the genus *Arctostaphylos*. This genus, the manzanitas and bearberries, consists of shrubs or small trees characterized by smooth orange or red bark and stiff, twisting branches. *Arctostaphylos* means "bear grapes," derived from Greek *arkto* (bear) and *staphyle* (grape). The fruits are small berries that ripen in the summer or autumn and are known to have provided an important food source for Native Americans. The pale pink, bell-shaped flowers bloom from March to May and are hermaphroditic, having both male and female organs. The plant is a burl-forming species, making it well adapted to fire. There are about 60 species of manzanita in the world. All of them are found in California, and most are found nowhere else.

Woollyleaf manzanita

Yerba santa

Yerba santa (*Eriodictyon californicum*), a member of the waterleaf family (*Hydrophyllaceae*), is a low, shrubby evergreen plant, two to four feet high, found growing on dry hills in California. A peculiar sticky resin covers the thick, leathery leaves. In spring, blueish flowers grow in clusters at the tips of the branches. It was given its name, "blessed herb," by Spanish priests impressed with its medicinal properties. Native Americans and early settlers used it for a cure-all. Yerba santa is regarded as good medicine for respiratory problems. A brew of its leaves serves as an expectorant that clears the breathing passages, and a poultice of the leaves can be used to treat pain in aching joints. If you crush the plant's leaves between your fingers, you will find them a bit sticky and aromatic.

California thrasher

The **California thrasher** (*Toxostoma redivivum*) has a charming personality and is captivating to watch. The California thrasher is a striking songbird with a thick, long, decurved, or downward-curved, bill. It feeds on the ground under the shelter of bushes, using its heavy bill to turn over leaf litter in search of food. No one can forget that beak: it looks like the tines of a hay thrasher, hence the name. A mimic, the California thrasher sings exuberantly year-round; its song has been likened to that of old world nightingales. Its beautiful song can be heard throughout the chaparral in California. The California thrasher commonly mimics a wide range of bird species that share its habitat. Vocal mimicry (one species being copied by a second species) is one way some species increase the size of their vocal repertoire. Why are sounds of other species sometimes incorporated into a bird's repertoire? The answer seems to be that an expanded repertoire may improve the ability to attract a mate or intimidate rivals.

A large resident sparrow, the **California towhee** (*Pipilo crissalis*) forages in the leaf litter by scratching with both feet at once, in a fast, hopping motion. This bird feeds on seeds and insects within the leaf litter or occasionally on berries or seeds in bushes. The California towhee has a uniform gray-brown body and a long tail with cinnamon-brown undertail feathers. The towhee builds a nest of thin twigs, stems with leaves, and flower heads, concealed in the dense foliage and lower branches of shrubs and trees. The nest is generally lined with fine stems, grasses, and hairs. Listen for the distinctive metallic *chink* next time you are wandering in maritime chaparral. The call of the California towhee is not very melodic, but it is unmistakable.

California towhee

The **wrentit** (*Chamaea fasciata*) is a small brown songbird with a long tail, stubby bill, and relatively uniform coloring. It is the only North American representative of the babbler family (*Timaliidae*), known for its continual and rapid vocalizations. Like other members in their taxonomic family, wrentits are very vocal. Their vibrant song is likened to the sound of a "bouncing ball" and is easily heard and recognized. It is a song that is familiar in any chaparral community. The song is a series of accelerating notes followed by a descending trill: *wren—tit—tit-tit-tit-t-t-t*. Wrentits skulk about in the dense chaparral and brushy coastal thickets. These quick-moving little birds are more often heard than seen.

Wrentit

The colorful, easy-to-recognize **buckeye butterfly** (*Junonia coenia*) has several distinctive wing spots that look like eyes. It has a single large eye spot on each of its upper forewings, and two eye spots, one large and one small, on its upper hindwings. The eye spots may be a way for this species to deter predators. However, when the buckeye's wings are closed, they are drab and easily overlooked. The underwings are camouflaged with browns and tans, allowing a resting butterfly to blend, undetected, into the environment. Males perch during the day on low plants or bare ground to watch for females, flying periodically to patrol or to chase other flying insects. Females lay eggs on the leaf buds and leaves of host plants. Buckeyes prefer open, sunny areas, where they rest on the bare ground, holding their wings out while basking.

Buckeye butterfly

Salinas River SB
Salinas R

**Salinas River—
National Wildlife Refuge**

**Marina Dunes—
Open Space Preserve**

Marina Dunes RV Park—

Marina Dunes Resort Hotel—

Marina State Beach—

**Locke Paddon—
Park**

**Monterey Bay Sanctuary—
Scenic Trail**

**South Monterey—
Bay Dunes**

BAY

MONTEREY BAY

1. DUNES DR
2. LAKE DR
3. CARMEL AVE
4. ZANETTA DR
5. REINDOLLAR AVE
6. HILLCREST AVE
7. BOSTICK AVE
8. SALINAS AVE

Marina

Marina
Skate Park

California State University
Monterey Bay

1. LIGHT FIGHTER DR
2. GIGLING RD
3. OWEN DURHAM ST

Fort Ord

Sand
City   Seaside

N

| 0 | 0.5 | 1 Mile |
| 0 | 0.5 | 1 Kilometer |

Marina State Beach

Salinas River to Sand City

	Sandy Beach	Rocky Shore	Trail	Visitor Center	Campground	Wildlife Viewing	Fishing or Boating	Facilities for Disabled	Food and Drink	Restrooms	Parking	Fee
Salinas River National Wildlife Refuge	•	•				•	•				•	
Marina Dunes Open Space Preserve	•	•				•						
Marina Dunes RV Park					•		•		•	•	•	•
Marina Dunes Resort Hotel	•	•							•		•	
Marina State Beach	•	•				•	•			•	•	
Locke Paddon Park		•				•	•			•	•	
Monterey Bay Sanctuary Scenic Trail	•	•					•					
South Monterey Bay Dunes	•	•										

SALINAS RIVER NATIONAL WILDLIFE REFUGE: *N.W. of Hwy. One, 3 mi. S. of Castroville.* This is an excellent spot for fishing, hiking, photography, and wildlife observation. Leave Hwy. One at Del Monte Blvd. (exit 412) and head north one mile on the dirt access road between artichoke fields; do not trespass. An unpaved parking area is the trailhead for a level mile-long path leading west and north to the ocean. The trail passes agricultural fields and thickets of coyote brush. Adjacent to the coastal dune ridge, the native vegetation includes species such as live-forever and colorful Indian paintbrush. The endangered Smith's blue butterfly feeds and nests on two species of buckwheat in the sand dunes. To protect plant and animal resources, stay on the trail. Avocets and black-necked stilts nest in the marsh north of the path, and red-necked phalaropes gather during migration. Caspian terns flock here in the summer, hoarsely crying what sounds for all the world like *maybe they* ARE...*maybe they* ARE.

The Salinas River mouth attracts California brown pelicans, gulls, and wintering ducks. The river flows from its headwaters in San Luis Obispo County, 170 miles southeast of this point. The Salinas River is unusual in that most of its water is out of sight, flowing beneath its bed. Once you reach the shore, you can hike several miles north along the strand into the neighboring Salinas River State Beach, except when storm waves breach the dune barrier and the river flows to the sea.

At the wildlife refuge, beach anglers take striped bass, starry flounder, sand sole, surfperch, steelhead, jacksmelt, and small sharks. Hunting for waterfowl is also allowed. No facilities; dogs and horses prohibited. Call: 510-792-0222.

MARINA DUNES OPEN SPACE PRESERVE: *N. end of Dunes Dr., Marina.* On the site of a former sand mine is a dune preserve revegetated with native plants. A one-third-mile-long gently sloping path leads from the end of Dunes Dr. to an overlook above the shoreline. The first part of the path is well packed, and the seaward end is loose sand. Stay on the path to protect the revegetation efforts. Typical dune species of the area include the Monterey spineflower, beach sagewort, and sand verbena. The black legless lizard may also be present in the dunes.

The sandy beach extends for miles in both directions, uninterrupted by natural barriers. To the north, private property separates the Marina Dunes from public lands near the mouth of the Salinas River. To the south, the beach is open past the former Fort Ord, where future public access improvements are planned. Dogs on leash are allowed in the Marina Dunes Open Space Preserve. Streetside parking; no facilities. Managed by the Monterey Peninsula Regional Park District; call: 831-372-3196.

MARINA DUNES RV PARK: *3330 Dunes Dr., Marina.* Use the Reservation Rd. exit off Hwy. One, and turn north on Dunes Dr., just before the Marina State Beach entrance. The private campground offers more than 60 RV sites, some wheelchair accessible, with patios, picnic tables, and barbecue grills. Individual and group tent campsites are also available. The park includes a playground, laundry, and gift shop. Pets allowed. Fee applies; reservations recommended. For information, call: 831-384-6914.

MARINA DUNES RESORT HOTEL: *3295 Dunes Dr., Marina.* A public path and boardwalk lead 500 feet over a dune to the steep sandy beach. Upon entering the hotel property, skirt to the right of the entrance building, and then turn left to park; walk seaward between the buildings to the beach path, which is open during daylight hours only.

MARINA STATE BEACH: *W. end of Reservation Rd., Marina.* Marina State Beach is located on high dunes, some of the tallest along Monterey Bay. Popular with picnickers, anglers, and surfers, the beach is not good for swimming due to strong currents. The elevated and breezy site makes this a popular spot for hang-gliders, paragliders, and model airplane enthusiasts.

The main park entrance on Reservation Rd. leads to a paved overlook with a fine view of the coast; parking, picnic tables, and bike racks are available. The beach is down the sandy slope from the parking area. Restrooms with running water are located east of the parking lot. Across the entrance road from the restrooms is a boardwalk, parts of which may be inundated by sand, leading south about one-half mile along the crest of the dunes. Markers describe typical plants and animals of the Monterey Bay dunes, including the endangered Menzies wallflower, blooming bright yellow in the summer. Dogs, horses, and fires are not allowed at Marina State Beach. For information, call: 831-384-7695.

Another accessway to Marina State Beach is located at the west end of Lake Dr. A loose sand path, one-third mile long, leads to the beach, up and over the 150-foot-high dunes. Loose sand makes this a more challenging access point, but the high elevation offers an attractive launch spot for hang-gliders. Pilots must first register at the Reservation Rd. parking lot and use appropriate safety equipment.

LOCKE PADDON PARK: *Reservation and Seaside Roads., Marina.* This park includes a freshwater pond that attracts waterfowl and is fed by rainfall and urban runoff. There is a wheelchair-accessible overlook and a pedestrian dirt path leading along the edge of the wetland. Picnic tables and restrooms are available, and nature walks and educational programs are offered. From the east side of

Marina Dunes Open Space Preserve

Marina State Beach

the park, bicyclists and hikers can access the Monterey Bay Sanctuary Scenic Trail, which runs along Del Monte Blvd. south to Monterey and Pacific Grove. Locke Paddon Park is maintained by the City of Marina; call: 831-884-1253. Improvements to the park are planned, in conjunction with construction of a new public library on an adjacent site.

MONTEREY BAY SANCTUARY SCENIC TRAIL:
Along Monterey Bay, from Santa Cruz to Pacific Grove. The Monterey Bay Sanctuary Scenic Trail, which incorporates variously named trail segments in different communities, is planned to ring Monterey Bay, connecting Santa Cruz and Monterey Counties. The trail offers hikers and bicyclists a safe off-road route to reach parks and beaches along the bay or to connect to other spur trails. A major purpose of the trail is to offer visitors views of the Monterey Bay National Marine Sanctuary, just offshore from the planned trail route. Much of the trail is within sight or sound of the sea; the segment through former Fort Ord provides the only trail access through the dunes until restored lands there become a future state park. Construction of a trail segment to link Aptos in Santa Cruz County and the City of Marina is planned by 2007. For information, call: 831-883-3750.

A half-mile east of the trail, which runs along Del Monte Blvd. near the Marina city hall, is the Marina Skate Park, 304 Hillcrest Ave. near Zanetta Dr. The park is free and open daily; helmets and safety pads required.

SOUTH MONTEREY BAY DUNES: *W. ends of Tioga Ave. and Bay St., Sand City.* Two public street ends provide shoulder parking and informal access to the beach. The Monterey Peninsula Regional Park District manages scattered park holdings in the dunes, mixed among undeveloped private parcels of land. The sandy beach extends south through Monterey Bay State Beach into the heart of Monterey. Call: 831-372-3196.

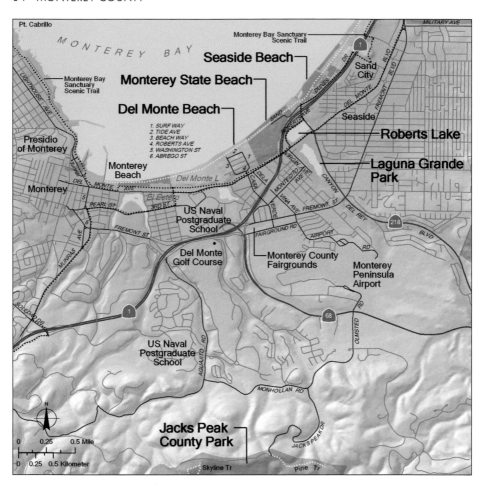

Seaside Beach

Seaside to Monterey

	Sandy Beach	Rocky Shore	Trail	Visitor Center	Campground	Wildlife Viewing	Fishing or Boating	Facilities for Disabled	Food and Drink	Restrooms	Parking	Fee
Seaside Beach	•	•					•			•	•	
Monterey State Beach	•	•					•			•	•	
Del Monte Beach	•	•					•			•	•	
Roberts Lake		•			•						•	
Laguna Grande Park		•				•	•			•	•	
Jacks Peak County Park		•				•	•			•	•	•

SEASIDE BEACH: *Sand Dunes Dr., W. of Hwy. One, Seaside.* A unit of Monterey State Beach is located on the north side of the Beach Resort Monterey. Paved parking, picnic tables, a sandy path to the beach, and boardwalks through a small area of vegetated dunes. Portable restrooms. Call: 831-649-2836.

MONTEREY STATE BEACH: *Sand Dunes Dr., W. of Hwy. One, Seaside.* This unit of Monterey State Beach, known as Houghton M. Roberts Beach, is located at the south end of Sand Dunes Dr. It has parking, restrooms, and access down the sloping dune to the sandy beach.

DEL MONTE BEACH: *N. of Del Monte Blvd., Monterey.* Turn off Del Monte Blvd. toward the bay at the stoplight at Casa Verde and bear right on Roberts Ave., then left on Surf Way to the beach. This park, maintained by the City of Monterey, offers broad vistas of Monterey Bay and the wooded Monterey Peninsula from its elevated site, which slopes gently down to the water's edge. A wheel-chair-accessible boardwalk leads through the dunes, and benches are placed to take advantage of the views. Restrooms located at the south end of Tide Ave. at Beach Way. Street parking; wheelchair-accessible parking at end of Beach Way.

Del Monte Beach

The California Department of Parks and Recreation is the state's premier supplier of coastal recreation. More than a quarter of the 1,100-mile California coastline, some 295 miles of shore, is encompassed within the state park system. The state park system draws 80 million visitors annually; of the ten most-visited California state park units, nine are sandy beaches.

On a coast with a rapidly growing population, state parks provide unparalleled natural scenery, recreation from sunset viewing to surfing the ultimate wave, and even solitude. Coastal plant and animal communities, including many that are rare statewide, are well represented in state park units. More than one-third of the species listed by the state and federal government as threatened or endangered can be found in the state park system. The protection of habitat is the primary goal for most of the land and water in state parks and reserves, but recreational opportunities are abundant. State marine parks include underwater areas that can be explored by divers.

Interpretive programs and campfire talks are offered at many locations. Thousands of volunteers help staff the parks; at Point Lobos State Reserve, docents are available to point out the dusky-footed wood rat nests tucked almost out of sight among the Monterey cypress trees. Most state park units protect cultural resources as well as natural resources. Chumash Painted Cave State Historic Park near Santa Barbara contains California's largest collection of Native American rock art. San Luis Obispo County's Hearst San Simeon State Historical Monument, known as Hearst Castle, is California's most popular state historical monument. On California's Central Coast are more than 30 other state park units, from Zmudowski State Beach to Point Mugu State Park. Many facilities are accessible to visitors with impaired mobility.

Affordable overnight accommodations overlooking the beach or set in a forest are provided by many state parks. For visitors arriving without a vehicle, hike or bike campsites have especially low fees; equestrian campsites are available in some locations. Environmental campsites offer simple facilities in natural settings, away from cars and crowds. Enroute camping, for those with a self-contained recreational vehicle, is available at many locations, and others offer trailer or RV hookups. Some campsites are available to those who arrive first, while others can be reserved up to seven months in advance; call: 1-800-444-7275.

California's State Parks Department supports local parks, too. Proposition 40, the California Clean Water, Clean Air, Safe Neighborhood Parks and Coastal Protection Act that was approved by voters in 2002, provided $870 million that is distributed to local park agencies and nonprofit groups for recreation facilities in all parts of the state.

The Golden Poppy Annual Day Use Parking Pass provides one year of unlimited access to 95 state park units, although not to certain locally managed or high-demand parks such as Hearst Castle. Other pass programs provide a discount for disabled persons, including military veterans, and for certain low-income persons. To learn more about pass programs, contact the State Parks Store at 1-800-777-0369 ext. #5. For information about state parks, see the department's website at www. parks.ca.gov, or call: 916-653-6995.

ROBERTS LAKE: *Roberts Ave. and Canyon Del Rey, Seaside.* Roberts Lake and Laguna Grande were once a single brackish lagoon with an outlet to Monterey Bay. Over the years, filling for development and transportation purposes divided the lake and cut off tidal action from the formerly estuarine complex. The reeds on the west side of Roberts Lake, opposite a small grassy picnic area, are habitat for green herons and pied-billed grebes. The lake is also a popular model boat racing area. Parking on Roberts Ave. The site adjoins the Monterey Bay Sanctuary Scenic Trail. Call: 831-899-6825.

LAGUNA GRANDE PARK: *Off Del Monte Ave., along Canyon Del Rey Blvd., Seaside, and Virgin Ave., Monterey.* Laguna Grande is a freshwater lake that is lined with tules and cattails, a habitat for red-winged blackbirds. A paved path around Laguna Grande crosses the lake via a bridge; grassy picnic areas have barbecue pits, and a playground has volleyball nets for rent. Restrooms and parking are available on both east and west sides. Blues concerts are offered on Sunday afternoons in late summer. The portion of the park along Canyon Del Rey Blvd. is maintained by the city of Seaside; call: 831-899-6825. On the other side of the lake, off Virgin Ave., the city of Monterey manages park facilities; call: 831-646-3866.

JACKS PEAK COUNTY PARK: *Jacks Peak Dr., Monterey.* Although located inland from the coast, the high ridge on which Jacks Peak County Park is located offers magnificent views of the ocean and Monterey Bay. Views are filtered by the native Monterey pine forest, which cloaks the peak.

There are eight miles of hiking and equestrian trails, picnic areas with tables and barbecue grills, and restrooms. The Skyline self-guided nature trail leads visitors through an area of Miocene-epoch fossils; descriptive brochures are available at the entrance station. Open daily; hours vary by season. Vehicle entrance fee. Managed by Monterey County Parks Department; for information, call: 888-588-2267. Jacks Peak County Park is named for Scottish immigrant David Jacks, whose local dairies produced "Monterey Jack" cheese, a California creation.

Del Monte Beach

Started in 1958, the Monterey Jazz Festival is the world's longest-running event of its kind. The festival takes place annually in September at the Monterey County Fairgrounds, off Hwy. 68 and Fairground Rd. The Next Generation Festival, featuring young jazz performers, takes place during the spring season at venues in downtown Monterey. For festival information, call: 831-373-3366.

In 1967 the Monterey County Fairgrounds was the site of the three-day Monterey Pop Festival, which featured such performers as Janis Joplin, Jefferson Airplane, and the Grateful Dead. The festival marked a turning point in rock music as it introduced Jimi Hendrix to America and was the nation's first huge outdoor rock concert.

Monterey Bay Dunes

D UNES have been a part of the Monterey bay landscape for at least the last million years. The Aromas sand, which is approximately one million years old, is
widespread east of Monterey Bay and is the geologic record of a sheet of sand
that was primarily blown here by the wind. This sandy geologic formation has been
overridden south of the Salinas River by a series of dunes, the oldest of which were
formed well before the latest period of rapid sea level rise, called the "Flandrian Transgression," roughly 18,000 to 6,000 years ago. These so-called pre-Flandrian dunes have
been stabilized by vegetation and are much more subdued in expression than their
younger counterparts, but are still quite recognizable. The depressions between dunes
collect water and commonly host wetlands. Soils generally are well developed on these
older sands and can be very productive—the Monterey Bay area is the artichoke capital
of the world.

The pre-Flandrian dunes are in turn overridden by more recent dunes that formed during and after the Flandrian Transgression. These younger dunes are scarcely stabilized
by vegetation, and are parabolic in shape, with the open end pointing seaward. They
are on the move, and particular dunes may come and go as the sand is blown away. As
the younger dunes are stripped away, the older, more resistant pre-Flandrian dunes are
left behind, protected by their soil crust.

Dunes of all ages are, however, susceptible to erosion by waves. As sea level has slowly
risen, marine erosion has carved away at the dunes, leaving a bluff that may reach
heights of as much as eighty feet. Because the loose sand of the dunes is so easily eroded, some of the greatest long-term erosion rates measured along the California shoreline occur in the southern Monterey Bay area. The eroding dunes provide a source of
sand to the marine environment. Much of the sand ends up back on the beaches, where
it is susceptible to being picked up once more by the wind, to form dunes once again.
Abrasion and weathering of the sand grains that occur in the beach environment serve
to remove the softer and less durable components, leaving young dunes of almost pure
quartz, which is resistant to weathering and abrasion. This pure quartz sand is a highly
attractive resource for the manufacture of glass. Sand of less pure quality is still eagerly
sought for use in making concrete. Mining of both commodities has been greatly cur-
tailed in the past decade due to concerns over the loss of dune habitat and sand supply
to the beach.

Artichoke field and Monterey Bay dunes

Monterey spineflower (*Chorizanthe pungens* var. *pungens*) is a soft, hairy, low-growing herb endemic to coastal areas in southern Santa Cruz and northern Monterey Counties. Monterey spineflower is a member of the buckwheat family (*Polygonaceae*) and is in a genus of wiry annual herbs that inhabit dry, sandy soils. Because of the patchy and limited distribution of such soils, many species of *Chorizanthe* tend to be very localized in their distributions. Portions of the coastal dune and coastal scrub communities that support Monterey spineflower have been eliminated or altered by recreational use, industrial and urban development, and military activities. This is one reason that this species is listed as threatened, pursuant to the Federal Endangered Species Act.

Monterey spineflower

Menzies wallflower (*Erysimum menziesii*), a low-growing succulent herb, is a member of the mustard family. This species produces dense clusters of bright yellow, four-petaled flowers from February to April. The characteristic fleshy, spoon-shaped rosette leaves distinguish this coastal species from other wallflowers. Menzies wallflower is restricted to foredunes and dune scrub communities. Due to habitat degradation by commercial and residential development, off-road vehicle use, trampling by hikers and equestrians, and sand mining, the Menzies wallflower is listed as endangered by the U.S. Fish and Wildlife Service and the California Fish and Game Commission. Look for this beautiful showy wildflower at Asilomar dunes, where the California Department of Parks and Recreation has restored historic dune habitat.

Menzies wallflower

Monterey manzanita (*Arctostaphylos montereyensis*) is a beautiful shrub species with evergreen leaves, slick reddish bark, and flowers that are tiny, sweet-smelling, and bell-shaped. Manzanita berries are an important food source for birds and small mammals. The Spanish called the tiny red fruit of this coastal shrub *manzanita*, meaning "little apple."

Monterey manzanita

Eastwood's goldenbush

Eastwood's goldenbush (*Ericameria fasciculata*), also known as mock heather, is a compact, evergreen shrub, one to three-and-a-half feet tall, in the *Asteraceae*, or sunflower, family. It is endemic to the Monterey County dunes and sandy areas near the coast. Eastwood's goldenbush has resinous herbage, peeling bark, and green leaves clustered on the stem. The radiating, pale yellow flowers sit in terminal clusters of two to six flowers. The plant blooms from August to November and is ranked by the California Native Plant Society as extremely rare. When you walk in the dunes in the late summer or fall and spy a yellow-flowering shrub, it may be Eastwood's goldenbush.

Seaside bird's beak

Seaside bird's beak (*Cordylanthus rigidus littoralis*), a California state-listed endangered plant species that grows in sandy soils and stabilized dunes, is a member of the figwort or *Scrophulariaceae* family. It is a bushy, annual herb with yellowish green branches and leaves that are covered with fine hairs. This unique plant can bloom under hot, harsh conditions when most annuals have ceased to grow. This is due to its ability to tap into the roots of neighboring plants and draw water and nutrients from its host to aid its own growth.

Smith's blue butterfly

Smith's blue butterfly (*Euphilotes enoptes smithi*) spends its entire life in association with two buckwheat plants in the genus *Eriogonum*. Emerging in late summer and early autumn, the adult butterflies mate and lay eggs on the flowers of these host plants. The eggs hatch shortly thereafter, and the larvae feed on the flowers of the plant. After several weeks of feeding and development, the larvae build cocoons, beginning a ten-month period of transformation. The next spring, when the buckwheat plants flower again, the adult butterflies emerge from the cocoons. Due to increasing automobile and foot traffic along the coast and the invasion of introduced plants, habitat for the Smith's blue butterfly has been degraded. This species was listed as endangered by the federal government in 1976.

Kangaroo rats are not related to the typical rat that most people imagine. These cute little rodents are well adapted for survival in arid environments. The **Salinas kangaroo rat** (*Dipodomys heermanni goldmani*) occurs in maritime chaparral habitat with sandy soils. The kangaroo rat has big hind feet and a long tail to help it balance when it jumps along at night, gathering seeds. The kangaroo rat places the seeds into fur-lined pouches inside its cheeks and later deposits them in a storage area called a *cache*. The short forefeet have strong claws for digging burrows in the sand. Kangaroo rats construct elaborate burrow systems where they spend the day out of the heat, protected from predators. When they are done foraging at night and return to their burrows, they push up a little plug of sand in front of the burrow opening to deter intruders. The next time that you take a hike in the Fort Ord open space, look for a small burrow hole that appears to have been filled in, and you may be lucky enough to find the home of a Salinas kangaroo rat.

Salinas kangaroo rat

The **black legless lizard** (*Anniella pulchra nigra*) is a burrowing reptile that looks like a snake and is about the size of a pencil. However, it is actually a lizard that has adapted to a legless condition to allow it to live a subterranean life in loose, sandy soil. The black legless lizard lives in the dune habitat where there is moisture, warmth, loose material to burrow in, and plant cover. The black legless lizard usually forages at the base of shrubs or other vegetation, either on the surface or just below it in leaf litter or sandy soil. It feeds mainly on insects, insect larvae, and spiders. The lizards sometimes seek cover under surface objects such as flat boards and rocks, where they lie barely covered. If you decide to look for legless lizards under a piece of wood or rock, please be sure to put the cover back in its original position.

Black legless lizard

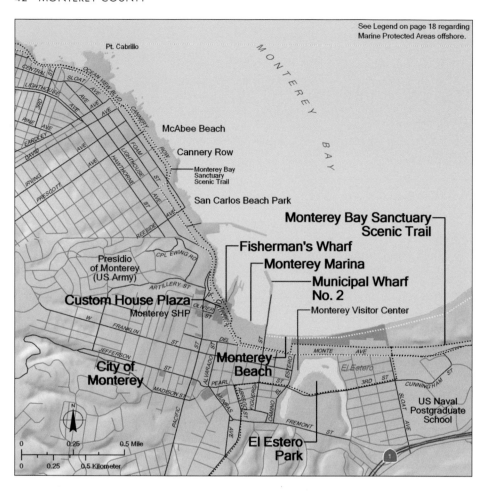

See Legend on page 18 regarding Marine Protected Areas offshore.

Pt. Cabrillo

MONTEREY BAY

McAbee Beach

Cannery Row

Monterey Bay Sanctuary Scenic Trail

San Carlos Beach Park

Monterey Bay Sanctuary Scenic Trail

Fisherman's Wharf

Monterey Marina

Municipal Wharf No. 2

Presidio of Monterey (US Army)

Custom House Plaza
Monterey SHP

Monterey Visitor Center

City of Monterey

Monterey Beach

El Estero

US Naval Postgraduate School

0   0.25   0.5 Mile

0   0.25   0.5 Kilometer

El Estero Park

Monterey Bay Sanctuary Scenic Trail

Monterey Central

	Sandy Beach	Rocky Shore	Trail	Visitor Center	Campground	Wildlife Viewing	Fishing or Boating	Facilities for Disabled	Food and Drink	Restrooms	Parking	Fee
Monterey Bay Sanctuary Scenic Trail	•	•	•			•		•	•	•	•	
El Estero Park			•			•	•	•	•	•	•	
Monterey Beach	•	•				•	•			•	•	
Municipal Wharf No. 2							•	•	•	•	•	•
Monterey Marina							•	•		•	•	•
Fisherman's Wharf						•	•	•	•	•	•	•
City of Monterey	•	•	•	•	•	•	•	•	•	•	•	
Custom House Plaza			•	•				•		•	•	

MONTEREY BAY SANCTUARY SCENIC TRAIL: Parallel to waterfront, Monterey.
A paved bicycle path and trail runs along the old Southern Pacific Railroad right-of-way, from Marina and Seaside to Lovers Point in Pacific Grove; a link to Santa Cruz County is planned. The wheelchair-accessible trail runs near many of Monterey's most popular attractions.

EL ESTERO PARK: Del Monte Ave. and Camino El Estero, Monterey.
A grassy city park inland of Del Monte Ave. features pedal boats on a lake and picnic areas. Dennis the Menace Playground contains a steam locomotive, a climbing wall, a pint-size maze, and other unusual play equipment. El Estero attracts ducks and other migratory birds; do not feed the birds. There is lake fishing from a pier off 3rd St.; no license required for anglers under 16. The El Estero skateboard park is located north of the ball field, next to the lake. For information, call: 831-646-3866.

MONTEREY BEACH: Shoreline E. of Municipal Wharf No. 2, Monterey.
A broad sandy beach popular with sunbathers and kayakers. Sand volleyball courts, picnic facilities, and the Monterey Bay Sanctuary Scenic Trail are adjacent to the beach. The city of Monterey has been gradually expanding the Monterey Bay Waterfront Park between the beach and Del Monte Ave. as part of a project known as the Window on the Bay. The park offers fine views of Monterey Bay. For information, call: 831-646-3866.

MUNICIPAL WHARF NO. 2: Foot of Figueroa St., Monterey.
An active commercial fishing operation is located at Municipal Wharf No. 2; sanddab, sole, squid, shrimp, salmon, rockfish, northern anchovy, and Pacific herring are unloaded here. Public facilities include restaurants, snack bar, and restrooms; metered parking. A three-ton capacity token-operated public boat hoist is available 24 hours a day; hoist use requires training. A 700-foot fishing promenade is on the east side of the wharf; no fishing license required for sport fishing. For information about the wharf, call: 831-646-3950.

MONTEREY MARINA: Between Fisherman's Wharf and Municipal Wharf No. 2, Monterey.
The Municipal Marina has 413 slips up to 50 feet in length. Visiting boats may rent slips on a no-reservation basis; call the Harbormaster: 831-646-3950. Two public launch

Monterey Beach

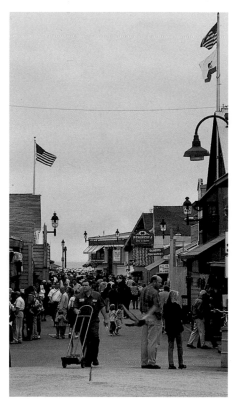

Fisherman's Wharf

Kayak rentals, instruction, and wetsuit rentals are available on Monterey Beach, east of Municipal Wharf No. 2.

Sailboat rentals, fishing boat charters, and wildlife-watching trips are available from several vendors on Fisherman's Wharf.

Several shops in the downtown area rent bicycles and other recreational equipment.

The Monterey Sports Center at 301 E. Franklin St. has a swimming pool with water slide, gymnasium, tennis courts, locker rooms, café, and pro shop; fee for facility use. Call: 831-646-3700.

ramps are located next to the Harbormaster's office; no fee required. Metered parking is available for vehicles with boat trailers. The Monterey Harbor and Marina are open daily, 24 hours a day.

FISHERMAN'S WHARF: *Foot of Alvarado St., Monterey.* A wharf was constructed on the site in 1870 by the Pacific Coast Steamship Company, which provided passenger and freight service along the west coast of North America. In 1913, the city took over the facility, which came to be called Fisherman's Wharf. Sardines were big business; on some days, 10,000 or more cases of the canned fish were shipped from the wharf. Visitor-oriented uses came to predominate after World War II, when the sardine fishery collapsed due to overfishing. Fisherman's Wharf features restaurants, gift shops, art galleries, and tackle and bait shops. Harbor cruises and fishing and whale-watching trips can be arranged on the wharf. Fee parking. Call: 831-649-6544.

CITY OF MONTEREY: *20 mi. S.W. of Salinas.* Monterey was California's political and social center until the gold rush of 1849, when San Francisco became the major port of trade. Once a port for New England hide and tallow traders, Monterey was also the site of a Portuguese shore whaling operation, a Chinese fishing village, and later, a world-renowned sardine canning industry. Tourism is the biggest industry now, and Monterey offers a wide range of visitor attractions.

CUSTOM HOUSE PLAZA: *Foot of Fisherman's Wharf, Monterey.* Custom House Plaza has been a focus of activity in Monterey since the city's early days; the Custom House was built in 1827. In the historic building now are the Pacific House Museum and the Monterey Museum of the American Indian. Open variable hours; call: 831-649-7118.

The Maritime Museum of Monterey, at 5 Custom House Plaza, holds the Allen Knight Collection of ship memorabilia and models. It has the Fresnel lens used for 90 years at Point Sur Lighthouse. Museum gift shop includes ship models. Open daily except Wednesday from 10 AM to 5 PM; for information, call: 831-372-2608.

In August 1879, Scottish writer Robert Louis Stevenson took up residence in a boarding house in Monterey. Stevenson, then in his mid-20s, came to town in order to be near Fanny Vandegrift Osbourne, a married woman with whom he had commenced an affair several years earlier. The couple had met in France, where she was studying painting and he was pursuing the Bohemian life of a writer, rather than the career in engineering or law that his father might have preferred.

His journey from Scotland to Monterey provided material for Stevenson's strong storytelling skill and acute powers of observation. *Treasure Island*, his adventure story involving a buried fortune and a band of murderous, if comically inept, buccaneers, hints at a California connection. The reader is given to suppose that the locale is somewhere off South America, the area known by British pirates as the Spanish Main. But consider the words of Jim Hawkins, the story's youthful protagonist: "I have never seen the sea quiet round Treasure Island. The sun might blaze overhead, the air be without a breath, the surface smooth and blue, but still these great rollers would be running along all the external coast, thundering and thundering by day and night; and I scarce believe there is one spot in the island where a man would be out of earshot of their noise." Then compare to excerpts from a travel essay by Stevenson about Monterey: "…from all round, even in quiet weather, the low, distant, thrilling roar of the Pacific hangs over the coast and the adjacent country like smoke above a battle." And—"…go where you will, you have but to pause and listen to hear the voice of the Pacific."

In *Treasure Island*, Jim Hawkins remarks upon rattlesnakes and sea lions, both characteristic of the Monterey Peninsula but not native to the Caribbean, or Scotland. He describes unfamiliar vegetation: "Then I came to a long thicket of these oaklike trees—live, or evergreen, oaks, I heard afterwards they should be called—which grew low along the sand like brambles, the boughs curiously twisted, the foliage compact, like thatch." Elsewhere, Stevenson writes about Monterey: "The crouching, hardy, live oaks flourish singly or in thickets—the kind of wood for murderers to crawl among..." In short, the island of the novel, published several years after Stevenson visited California, sounds less like the palmy, tropical isles of the Caribbean than the pine-forested hills of Monterey.

Following her divorce, Fanny married Robert in 1880, and they settled eventually on the Pacific isle of Upolu, in Samoa. There Robert continued to write, and there he died in 1894. His grave lies on a luxuriantly forested promontory, above the little harbor town of Apia, facing the Monterey Peninsula across the wide Pacific Ocean. The early 19th-century adobe house where Stevenson stayed is at 530 Houston Street, part of the Monterey State Historic Park; call: 831-649-7118.

Stevenson House

Monterey Bay Aquarium

TANKS of shimmering anchovies, ethereal jellies, and shape-changing octopuses provide a glimpse into the world within Monterey Bay and the sea beyond. The richness of life in the ocean along California's Central Coast is on display at the Monterey Bay Aquarium. The fascination of learning about plants and animals that few will ever see in their native habitat is matched, for many of us, by astonishment at the brilliant colors and unexpected forms of the organisms on display.

When it opened in 1984, the Monterey Bay Aquarium was an instant hit. With nearly 2.4 million visitors in its first full year, one million of them in the first five months alone, the aquarium set attendance records for United States aquariums. Founders Lucile and David Packard provided funds for initial construction, and subsequent expansion projects have made room for many more displays. The aquarium occupies a site on Monterey's Cannery Row on which a major fish cannery once stood. Opened in 1916 by Knut Hovden, the cannery processed sardines here until 1973. In the peak years of the 1940s, the sardine industry in Monterey processed 250,000 tons of fish each year.

The Monterey Bay Aquarium is a world leader in marine education and conservation. The aquarium has won awards for its exhibits and has many innovations to its credit, including the first living kelp forest display in an aquarium. Unlike many old-style aquariums, the Monterey Bay Aquarium displays fish, invertebrates, plants, and other organisms in contexts that match their origin, rather than divided by species among separate tanks. The million-gallon Outer Bay exhibit represents the largest and most diverse open-ocean exhibit found in any of the world's aquariums. On display are such large and striking animals as scalloped hammerhead sharks, mahi-mahi, bluefin and yellowfin tuna, Galapagos sharks, and black sea turtles. A thick sheet of acrylic plastic measuring 15 by 54 feet—the largest window in the world when installed in 1996—provides a clear view of ocean resources that are typically found in California's offshore waters. The companion Ocean's Edge exhibits focus on the coastal habitats of California's Central Coast, much of it easily accessible to visitors.

Cylinder of swimming anchovies

Jellies: Living Art exhibit

There are about 550 species on display at the Monterey Bay Aquarium, in over 100 display tanks. A constant stream of sea water is pumped through the aquarium's displays, filtered during the day to keep the exhibits as clear as possible, and unfiltered at night to provide the plankton and larvae that are food sources for many of the aquarium's inhabitants and that colonize the exhibit rockwork, contributing to its natural appearance. The water is kept at temperatures that match those of the habitats of the species on display. That favorable conditions are maintained is obvious; in 1998, an ocean sunfish (*Mola mola*) had to be returned to the wild after it outgrew its exhibit space, going in little more than one year from 57 to 880 pounds.

Among the most beautiful of the aquarium's creatures are the gently floating moon jellies. Southern sea otters, with a population in the wild that numbers fewer than 3,000 animals, are among the aquarium's most popular residents. Most are pups that were rescued after becoming accidentally separated from their mothers in the wild. Sea ot-

ters at the Monterey Bay Aquarium have the appearance, and sometimes the behavior, of furry rascals: they have unscrewed or unbolted parts of their exhibit, scratched the acrylic windows with shells, and tucked toys in water pipes. The aquarium's Sea Otter Research and Conservation program has rehabilitated injured sea otters and returned them to the wild.

As extensive as the aquarium's collections are, some notable species are not included. Whales, dolphins, and sea lions may be seen outside the aquarium, often not far from its ocean-view decks overlooking Monterey Bay.

The aquarium's very informative website draws three times as many visits as the aquarium itself;

Kelp Forest exhibit

Sea pens and tube anemones in the Sandy Seafloor gallery of Ocean's Edge wing

see: www.montereybayaquarium.org. The website includes an on-line field guide with photos and information on hundreds of species that inhabit the deep sea, the sea floor, coastal wetlands, rocky shore, and other habitats. The aquarium's popular Seafood Watch program provides recommendations to diners about the sustainability of various market fish. Video cameras at several of the aquarium's exhibits stream live action that is viewable on the website. The sea otter "webcam" is accompanied by a guide to help viewers note the differences among the individual sea otters on display. Other video cameras show sharks, the aviary, penguins, the Outer Bay exhibit, the Kelp Forest exhibit, and Monterey Bay itself. Learning resources available on the aquarium's website include activities that families and schoolchildren can do in conjunction with a visit, or in place of a visit. The very popular Underwater Explorers program offers aquarium visitors ages 8 to 13 an opportunity to accompany staff on a "dive," while using scuba equipment, in the aquarium tide pool. In addition to paid staff, the nonprofit aquarium relies on the efforts of more than 1,000 volunteers.

The aquarium conducts research in conjunction with Stanford University's Hopkins Marine Station. Research subjects include tuna in the world's oceans and white sharks, which inhabit the waters off California as well as many other places around the world.

The aquarium's independent sister institution, the Monterey Bay Aquarium Research Institute (www.mbari.org), was founded in 1987 by David Packard with a mission to carry on a range of research projects related to ocean science. The institute's studies

Monterey Bay Aquarium

address marine biology, the geology of the deep sea, climate studies, and new techniques for observing conditions in the deep ocean, among other topics. One research project has investigated the effects of increasing carbon dioxide in the atmosphere, while another has studied the dynamics of the coastal upwelling phenomenon along California's coast that brings nutrients from the deep sea to the surface. The Monterey Bay Aquarium Research Institute is located at Moss Landing.

The Monterey Bay Aquarium is open daily except Christmas Day, from 9:30 AM to 6 PM during the summer months and on holidays, and from 10 AM to 6 PM the rest of the year. Fee for entry. Same-day tickets are available at the aquarium; advance tickets are recommended during the summer and on holidays. Call the 24-hour information line at 831-648-4888; for information in Spanish, call: 1-800-555-3656. A free trolley provided by Monterey-Salinas Transit links the Monterey Bay Aquarium to Fisherman's Wharf and downtown Monterey during the summertime and on holidays; a second summertime trolley route connects the Aquarium with downtown Pacific Grove. For information, call: 831-899-2555.

Volunteer guide and visitors at touchpool in Ocean's Edge wing

See Legend on page 18 regarding Marine Protected Areas offshore.

Berwick Park

Pt. Cabrillo

MONTEREY BAY

Monterey Bay Aquarium

McAbee Beach

Cannery Row

Monterey Bay Sanctuary Scenic Trail

Hostelling International, Monterey

Aeneas Beach

San Carlos Beach Park

Coast Guard Pier

Breakwater Cove Marina

Fisherman's Shoreline Park

Presidio Museum of Monterey

CPL EWING RD

ARTILLERY ST

Fisherman's Wharf

Monterey Bay Sanctuary Scenic Trail

Monterey SHP

Monterey Marina

Monterey Beach

Veteran's Park

El Estero

US Naval Postgraduate School

Huckleberry Hill Nature Preserve

N

El Estero Park

0   0.25   0.5 Mile

0   0.25   0.5 Kilometer

Breakwater Cove Marina

Monterey and Cannery Row

	Sandy Beach	Rocky Shore	Trail	Visitor Center	Campground	Wildlife Viewing	Fishing or Boating	Facilities for Disabled	Food and Drink	Restrooms	Parking	Fee
Fisherman's Shoreline Park			•			•	•					
Breakwater Cove Marina							•	•	•	•	•	•
Coast Guard Pier						•	•	•		•	•	•
San Carlos Beach Park	•					•	•			•	•	
Aeneas Beach	•											
Cannery Row	•	•	•				•		•	•	•	•
McAbee Beach	•											
Monterey Bay Aquarium				•		•	•		•	•		•
Hostelling International, Monterey							•		•	•	•	•
Presidio Museum of Monterey				•			•		•	•		
Veteran's Park					•		•		•	•	•	
Huckleberry Hill Nature Preserve			•		•							

FISHERMAN'S SHORELINE PARK: *Between the Coast Guard Pier and Fisherman's Wharf, Monterey.* Benches and viewpoints are located along a narrow coastline park that incorporates part of the Monterey Bay Sanctuary Scenic Trail. Look for harbor seals and sea lions in the harbor waters.

BREAKWATER COVE MARINA: *32 Cannery Row, Monterey.* Located on the downtown side of the Coast Guard Pier is a marina with 80 slips, a boat hoist, boat storage, and boat repair facility. Guest boat slips available. There is a restaurant and deli overlooking the water; diving equipment and kayak rentals are also found here. Call for marina information, Monday to Friday, 8 AM to 12 PM and 1 to 5 PM: 831-373-7857. Fuel dock open every day; call: 831-647-9402.

COAST GUARD PIER: *S.E. end of Wave St., Monterey.* The Coast Guard Pier is part wharf and part breakwater that shelters the Monterey Marina. The pier is home port for Coast Guard vessels; public access is permitted from sunrise to sunset. Fishing is permitted on the north side of the breakwater. At the foot of the breakwater is a launch ramp for trailered boats and other small craft; a boat hoist for larger boats is available at the ad-

jacent Breakwater Cove Marina. Restrooms and showers are located at the foot of the breakwater. Facilities open daily. A change machine at the entrance provides coins for the parking meters. The rock jetty extension of the 1,700-foot-long breakwater serves as a haul-out for hundreds of seals and noisy sea lions. Adjacent San Carlos Beach provides a popular sandy beach entry point for divers. ·

SAN CARLOS BEACH PARK: *Foot of Reeside Ave., Monterey.* At the east end of Cannery Row is a pleasant park with green lawns, picnic tables, and access to the sandy beach by stair and ramp. Restrooms and an out-

Monterey Bay Sanctuary Scenic Trail

Native Americans had been living on the Monterey Peninsula for thousands of years when Portuguese explorer Juan Rodríguez Cabrillo, working in the service of Spain, spotted it while sailing along the coast in 1542. In 1602, Sebastián Vizcaíno was looking for a port suitable for Spanish galleons from Manila and named the harbor after the Count of Monte Rey, the viceroy of Mexico. The Spanish claimed sovereignty over the area in 1770, and seven years later, Monterey became the capital of the province of Las Californias. When the northern area was divided from Baja California in 1804, Monterey was made the capital of Alta California, a vast region that now comprises five western states.

The period under the Spanish was marked by the establishment of a presidio and a mission, soon moved to the present site in Carmel. The conversion of American Indians to Christianity also took place, and, unfortunately, so did the spread of deadly diseases among the indigenous people. After Mexico declared independence from Spain, Monterey remained the capital of Alta California. The Mexican period that began in 1822 brought a great deal of trade, commerce, and growth to Monterey, which was California's only international port.

In June of 1846, rebels in Sonoma led the Bear Flag Revolt against Mexican governance, but their self-declared "California Republic" ended a month later; they joined forces with the American military instead after Commodore Sloat came ashore—in Monterey—and claimed Alta California for the United States. When the Mexican-American War ended in 1848, Monterey and the rest of Alta California officially became part of the United States.

Before California achieved statehood, Monterey hosted the Constitutional Convention of 1849. Delegates established the Sierra Nevada range as California's eastern boundary, decided not to allow slavery, and gave women the right to own property. The delegates named San Jose as the first state capital, and Monterey's role as a capital city came to an end.

Historic buildings in Monterey include Larkin House, Cooper Molera Adobe, and Casa Soberanes. Monterey is home to California's first theater (still in use); its first newspaper (*The Californian*); its first brick building (built by a Donner Party survivor); and its first public library (established in 1849). California State Historical Monument #1, the Custom House, is part of Monterey State Historic Park, a focal point for many phases of California history.

door shower are available. Metered parking is adjacent. A popular beach and entry point for divers. At the west end of the beach, there is a public viewing deck overlooking the sea at the Monterey Bay Inn.

AENEAS BEACH: *400 Cannery Row, Monterey*. This sandy beach with rocky areas and tidepools can be reached by a stairway at the east end of the Monterey Plaza Hotel. A large public terrace, part of the hotel development, overlooks the beach, where div-

ers sometimes enter the water. There is also access to a rocky promontory viewpoint via a walkway under the Chart House Restaurant, on the west side of the hotel. A public walkway, located on the ocean side of the hotel, connects the beach on the east with the viewpoint on the west.

CANNERY ROW: *Between David Ave. and Coast Guard Pier, Monterey*. Originally the site of the largest Chinese fishing village on Monterey Bay, the area eventually became the center of a thriving sardine canning industry pioneered by Frank Booth, who built the first fish-packing plant near Fisherman's Wharf in 1902. In 1905 Knut Hovden from Norway modernized packing methods and developed a system for steam-cooking sardines that revolutionized the industry. Pietro Ferrante from Sicily introduced the lampara net in 1911, which greatly increased the sardine catch. Sardines were canned almost exclusively for human consumption until the end of World War I, when the price fell so low that an additional use was developed by Booth. Sardine by-products, previously considered waste, were reduced to fish meal for animal feed, and oil for manufacturing soap, tires, paint, vitamins, and glycerine. As a result of this more efficient use of the fish, and the introduction of the purse seiner boat in 1928, the annual sardine catch jumped from 3,000 tons in 1916 to 250,000 tons in 1945.

The sardine industry peaked in 1945—the same year that John Steinbeck's novel *Can-*

Kayak rentals on Cannery Row: A B Seas Kayaks, 831-647-0191.

Adventures by the Sea rents kayaks and bicycles: 831-372-1807.

nery Row was published; at that time 23 canneries and 19 reduction plants occupied nearly one mile of shoreline. The "silver harvest" continued, despite warnings of overfishing, until the sardines virtually disappeared in 1951. Why the Pacific sardine fishery collapsed is not fully understood; biologists believe it may have been due to overfishing during a down cycle. A resurgence of the sardines may be occurring.

MCABEE BEACH: *Foot of Prescott Ave., Monterey.* The sandy beach where divers enter the water is reached by a stairway from Steinbeck Plaza and at either side of Spindrift Inn. Spindrift Inn provides a public outside shower on the bay side of the building.

McAbee Beach

Aeneas Beach

MONTEREY BAY AQUARIUM: *Cannery Row at David Ave., Monterey.* The Monterey Bay Aquarium offers a unique introduction to the marine ecosystems of the bay and ocean. The internationally acclaimed aquarium contains numerous habitat displays, touch tanks, hands-on exhibits, an outdoor plaza overlooking Monterey Bay, a restaurant, a gift shop, and more. Fully wheelchair accessible. The aquarium is a nonprofit organization supported by entry fees and member contributions. Open daily except Christmas; for information, call the 24-hour information line at 831-648-4888. A free public viewpoint overlooking the bay is located next to the aquarium entrance, through the middle of the building. A marine protected area designation by the California Department of Fish and Game provides special protection to marine resources in state waters around Point Cabrillo.

HOSTELLING INTERNATIONAL, MONTEREY: *778 Hawthorne St., Monterey.* The hostel is located in a renovated historic carpenter's union hall, four blocks from Cannery Row. Self-serve kitchen; pancake breakfast in-cluded. Groups welcome; youth discounts. Bicycle rentals and Internet access available; wheelchair-accessible facilities; free parking. Open 8 to 10 AM; 5 to 10 PM; 11 PM curfew. For information, call: 831-649-0375.

PRESIDIO MUSEUM OF MONTEREY: *Corporal Ewing Rd. near Artillery St., Presidio of Monterey.* The city of Monterey manages a 26-acre park on the grounds of the historic Presidio of Monterey. The original Presidio, whose ruins are located downtown off Church and Figueroa Streets, was built in 1770; it was one of four major Spanish military forts built in California between 1769 and 1797. El Castillo was a fortification built in 1792 on a hill overlooking Monterey Bay to protect the Presidio. The present Presidio was built by the U.S. Army in 1902 on the site of El Castillo.

The military history of the area is the focus of the Presidio Museum of Monterey, located on Corporal Ewing Rd. near Artillery St. and operated by the city of Monterey. Open Monday, 10 AM to 1 PM; Thursday to Saturday, 10 AM to 4 PM; and Sunday, 1 to 4 PM. Free admission. Call: 831-646-3456.

VETERAN'S PARK: *W. end of Jefferson St., Monterey.* This delightful city park set high on a hill above downtown Monterey offers picnic areas and children's play equipment set among expansive lawns and a 40-site campground for tents or RVs up to 21 feet long. Hike or bike campers welcome. One campsite is wheelchair accessible. Restrooms, showers, and RV dump station available. Family campsites are first-come, first-served; youth group campsites and group picnic areas can be reserved in advance. Fee for camping. A park attendant resides in the park; call: 831-646-3865.

HUCKLEBERRY HILL NATURE PRESERVE: *Presidio of Monterey, W. of Veteran's Park.* Enter this city-run open space area through adjacent Veteran's Park. The 80-acre open space area, leased by the city of Monterey from the U.S. Army, contains hiking trails through Monterey pine forest and maritime chaparral. Fine filtered views of Monterey Bay through the trees. Call: 831-646-3865.

Monterey Bay Aquarium

Monterey Bay Aquarium, Kelp Forest exhibit

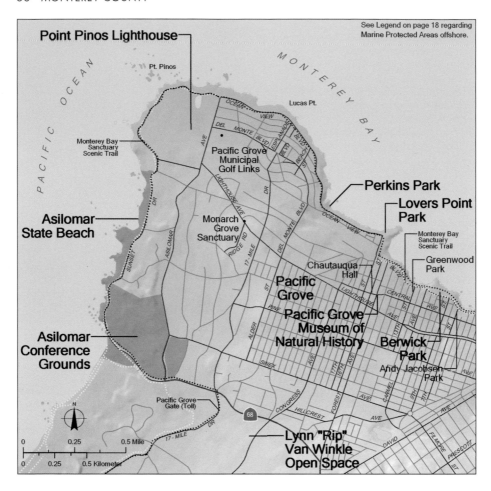

See Legend on page 18 regarding Marine Protected Areas offshore.

PACIFIC OCEAN

MONTEREY BAY

Point Pinos Lighthouse

Pt. Pinos

Lucas Pt.

OCEAN VIEW

DEL MONTE BLVD

AVE

ESPLANADE

BEACH ST

Monterey Bay Sanctuary Scenic Trail

Pacific Grove Municipal Golf Links

DR

DEL MONTE BLVD

OCEAN VIEW

Perkins Park

Lovers Point Park

Asilomar State Beach

Monarch Grove Sanctuary

SUNSET DR

ASILOMAR AVE

RIDGE RD

17 - MILE DR

Monterey Bay Sanctuary Scenic Trail

Chautauqua Hall

ST

LIGHTHOUSE AVE

CENTRAL AVE

ST

BLVD

Greenwood Park

Pacific Grove

Pacific Grove Museum of Natural History

PINE

ALDER

SINEX

AVE

16TH

17TH

AVE

FOREST AVE

CARMEL AVE

11TH ST

9TH ST

7TH

Berwick Park

Andy Jacobsen Park

AVE

Asilomar Conference Grounds

Pacific Grove Gate (Toll)

N

0   0.25   0.5 Mile

0   0.25   0.5 Kilometer

17 - MILE DR

CONGRESS

HILLCREST

68

AVE

DAVID

PRESCOTT ST

FILMORE ST

Lynn "Rip" Van Winkle Open Space

Berwick Park

Pacific Grove

	Sandy Beach	Rocky Shore	Trail	Visitor Center	Campground	Wildlife Viewing	Fishing or Boating	Facilities for Disabled	Food and Drink	Restrooms	Parking	Fee
Pacific Grove	•	•	•	•		•	•	•	•	•	•	
Berwick Park		•	•									
Lovers Point Park	•	•	•				•	•	•	•	•	
Perkins Park	•	•	•							•		
Pacific Grove Museum of Natural History				•				•	•	•		
Point Pinos Lighthouse				•						•		
Asilomar State Beach	•	•	•		•			•	•	•		
Asilomar Conference Grounds				•				•	•	•	•	•
Lynn "Rip" Van Winkle Open Space				•						•		

PACIFIC GROVE: *N.W. of the city of Monterey.* This residential community was founded as a Methodist retreat and tent city in 1875 and incorporated in 1889. The first Chautauqua in the West was organized here in 1879—a nationwide summer educational program with lectures and entertainment; Chautauqua Hall at 16th St. and Central Ave. is a State Historical Landmark and is still used for dances and other events. A Victorian Home Tour is held each October; call: 831-373-3304. The Pacific Grove Municipal Golf Links, at 77 Asilomar Ave., is an oceanfront 18-hole course, with white water views, pro shop, restaurant, and more affordable fees than other Monterey Peninsula courses; for information, call: 831-648-5777.

Monarch butterfly

Beginning in October, hundreds of thousands of monarch butterflies overwinter in Pacific Grove in native Monterey pine trees and introduced eucalyptus trees, collectively called "butterfly trees." Kindergarten children, dressed as monarchs, welcome the returning butterflies each fall with a parade in the town; call: 831-646-6540. Monarch viewing areas include a city-owned sanctuary off Ridge Rd. near Lighthouse Ave. and George Washington Park at Alder St. and Pine Ave.; the park also has a picnic area, restrooms, children's play facilities, and a baseball field.

The community of Pacific Grove is proud of its scenic shoreline. A series of city parks follow the blufftop above the town's rocky coast. The paved Monterey Bay Sanctuary Scenic Trail, serving bicyclists and pedestrians, connects sites around Monterey Bay and runs through Pacific Grove as far west as Lovers Point. From there to Point Pinos and beyond, the shoreline is publicly owned and bordered by a shorefront path. Stairs at several locations allow access to Pacific Grove's pocket beaches. Offshore waters are part of the Monterey Bay National Marine Sanctuary. Blue-jacketed volunteer docents from Bay Net, a Sanctuary affiliate, are often present along the Monterey Bay Sanctuary Scenic Trail, where they provide information to visitors about marine resources.

Pacific Grove

BERWICK PARK: *Along Ocean View Blvd. between Carmel Ave. and 9th St., Pacific Grove.* In 1542, Juan Rodríguez Cabrillo may have landed at the nearby point, now named for him. Part green lawn and part native plant landscaping, this small city park offers dramatic vistas of Monterey Bay and Pacific Grove's rocky shoreline. A paved pedestrian and bicycle path extends through the park, with benches scattered along it. Street parking. Two nearby city parks, both located on the inland side of Ocean View Blvd., offer additional open space and views: Andy Jacobsen Park is at the foot of 7th St. and Greenwood Park is at the foot of 13th St.

LOVERS POINT PARK: *End of 17th St. at Ocean View Blvd., Pacific Grove.* A grassy blufftop park dotted with Monterey cypress trees is situated on this rocky point. Paths along a seawall overlook the water. A short concrete pier structure on the east side of the point separates two small, protected sandy beaches, accessible by stairs. These beaches provide one of the few safe swimming areas along Monterey County's shoreline and are extremely popular. Other park activities include picnicking, sunning, fishing, surfing, and diving. Facilities include benches, picnic tables, barbecue grills, and restrooms. A sand volleyball court and children's swimming pool are also provided, and there is a restaurant and snack bar. Kayak and bicycle rentals available in the park. Parking at the corner of Ocean View Blvd. and 17th St. The park is managed by the city of Pacific Grove; call: 831-648-5730.

PERKINS PARK: *Along Ocean View Blvd., N.W. of Lovers Point, Pacific Grove.* Perkins Park occupies the shoreline to the west of Lovers Point. Paths wind through the colorful, but nonnative, ice plant. Benches are sited to take in the wonderful views of surf, rocky coast, and bay. Mostly street parking; there is a small parking lot at the foot of Beach St. and another at the foot of Asilomar Ave., where there are picnic tables and grills. Four stairways provide access to small pocket beaches. Undeveloped Esplanade Park is located at Esplanade and Ocean View Boulevards.

PACIFIC GROVE MUSEUM OF NATURAL HISTORY: *Forest Ave. at Central Ave., Pacific Grove.* The museum features the natural history of the Monterey area, with exhibits on native plants and animals, local geology, and the history of the area's indigenous peoples. The annual Wildflower Show, held in April, displays hundreds of species in bloom, identified and labeled by members of the California Native Plant Society. The museum includes a volunteer-staffed gift shop. For information, call: 831-648-5716. Open Tuesday through Saturday, from 10 AM to 5 PM, except major holidays.

POINT PINOS LIGHTHOUSE: *Asilomar Ave., S. of Ocean View Blvd., Pacific Grove.* Point Pinos, or Pine Point, is the rocky, low-lying northerly tip of the Monterey Peninsula. Sea stars, snails, black abalone, and marine algae inhabit the tidepools; whales and pelagic birds can be seen from the point. A marine protected area designation by the California Department of Fish and Game provides special protection to marine resources in state waters around Point Pinos. Offshore rocks, islets, and exposed reefs off Point Pinos and elsewhere along the state's coast are part of the California Coastal National Monument.

Point Pinos Lighthouse, built in 1885, is the oldest continuously operating lighthouse on the West Coast. The Fresnel lens of the third order, meaning of medium size and focal length, was manufactured in France in 1853. At first, the light source was a lantern fueled by whale oil, followed by lard oil and later kerosene. The lamp is now a 1,000-watt electric bulb that produces a beacon visible up to 15 miles at sea. Operated by the Pacific Grove Museum of Natural History; open Thursday through Monday, from 1 to 4 PM; during the summer, open daily, 11:30 AM to 5 PM. Donation requested; for information, call: 831-648-5716 ext. 13.

ASILOMAR STATE BEACH: *W. of Sunset Dr., Pacific Grove.* Asilomar State Beach includes rugged rocky shore, tidepools, sandy beach, and dunes. Inland of the dunes is Monterey pine forest; look for acorn woodpeckers among the trees. Past practices of livestock grazing and uncontrolled public access caused significant damage to the sand dune environment. The area has been revegetated by the Department of Parks and Recreation, and a wheelchair-accessible boardwalk has been installed to lead visitors through the dunes without damage to the habitat.

On Asilomar Beach, the distinctive white sand comes from local granodiorite rocks, which are weathered and broken down by the surf. A free beach wheelchair for use on the sand and boardwalk may be reserved in advance; call: 831-372-4076. Do not disturb plants or creatures in the tidepools. No fires on the beach; dogs must be on a six-foot leash. Parking areas are spaced along Sunset Dr.

ASILOMAR CONFERENCE GROUNDS: *Asilomar Ave. at Sunset Dr., Pacific Grove.* Asilomar means "refuge by the sea," and this state park conference center is a popular retreat destination. The facility began as a Young Women's Christian Association (YWCA) camp, with Arts and Crafts–style structures designed by Julia Morgan, a renowned California architect and designer of Hearst Castle. In 1951, the property became a unit of the state park system. Meeting rooms and overnight lodging facilities are available for group events; guest rooms and meals are also available to individuals. Guest rooms are simple, but have private baths; some have fireplaces. For individual reservations, call: 831-642-4242.

LYNN "RIP" VAN WINKLE OPEN SPACE: *Congress Ave. between Sunset Dr. and Forest Lodge Rd., Pacific Grove.* This open space park contains native Monterey pine, Monterey cypress, and coast live oak trees. Walking paths through the trees; dogs allowed off-leash. Street parking. From the Open Space, there is free pedestrian access to the Del Monte Forest and 17-Mile Dr. via the Country Club Gate.

Point Pinos Lighthouse

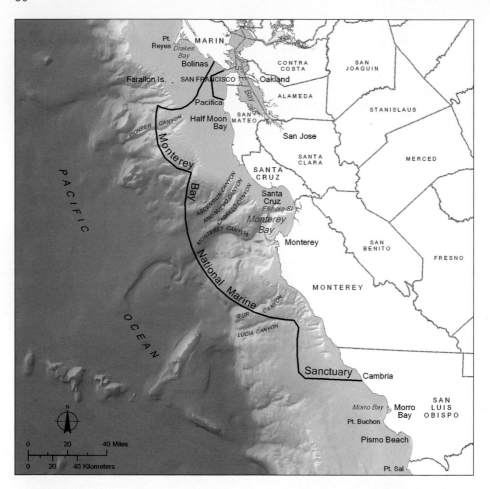

Sea otters, Monterey Bay National Marine Sanctuary

Monterey Bay National Marine Sanctuary

T HE DIVERSITY and abundance of natural resources along California's Central
Coast extends both on and offshore. Forests, canyons, peaks, and plains are found
on the land and also under the sea. Bathed by the chilly waters of the Pacific Ocean
and populated by hundreds of kinds of fish, invertebrate animals, and plants, the seafloor is close at hand, but out of sight to all but divers. The Monterey Bay National
Marine Sanctuary was created to protect the marine resources of the area centered on
Monterey Bay and its colossal undersea canyon, and to provide opportunities for study
and education. The Marine Sanctuary extends from southern Marin County to Cambria in San Luis Obispo County, and from the mean high-water line along the shore
seaward an average of 30 miles. The Monterey Bay National Marine Sanctuary is the
largest of the nation's marine sanctuaries, encompassing over 5,300 square miles, and it
is inhabited by 33 species of marine mammals, 94 species of seabirds, and nearly countless numbers of other organisms.

Despite the word *sanctuary* in its title, the Monterey Bay National Marine Sanctuary is
heavily used by humans. Commercial and recreational fishing are allowed, and cruise
ships, freighters, and tankers pass through or near the Sanctuary. Swimming, surfing,
boating, kayaking, and diving, along with other recreational uses, are permitted within
the Sanctuary, although the use of motorized personal watercraft is restricted to four
designated locations. When visiting the shoreline of the Marine Sanctuary or its waters,
leave tidepool creatures and other animals as you find them, and dispose of trash properly. Collection of rocks and shells within the Sanctuary's boundaries is not allowed,
with limited exceptions, one of which is the collection of jade at Jade Cove, south of
Sand Dollar Beach in Monterey County.

The Marine Sanctuary offers opportunities to learn more about the ocean, its inhabitants, and their relationships with life on land. Along the border of the Sanctuary are
over 30 research institutions, including the University of California at Santa Cruz, California State University Monterey Bay, the Moss Landing Marine Laboratory, the Monterey Bay Aquarium Research Institute, Stanford University's Hopkins Marine Station,
and the Monterey Bay National Marine Sanctuary itself. Area institutions share data
and cooperate in many ways in a wide variety of studies of the marine environment. As
vast as they are, the world's oceans are not the infinite resources they were once imag-
ined to be, where the bounty of fish to be taken seemed inexhaustible and where places
to dump the discards of human civilization seemed limitless. As the fallacy of these as-
sumptions has become apparent, so has the importance of learning more about the role
the world's oceans play in climate, the world's ecosystems, and human welfare.

One of the research programs conducted in the waters of the Monterey Bay National
Marine Sanctuary involves the market squid. Known to diners as calamari when fried
in crispy chunks or tossed with pasta, California market squid (*Loligo opalescens*) are an
ecologically important marine species in the waters off California. The market squid
fishery is the state's largest, although much of that catch is exported, while California
diners typically eat squid imported from Asia. Studies conducted in 2004 and 2005 in
Monterey Bay tested the use of sonar, an acoustic technology, to measure the abundance of market squid eggs deposited on the sandy seafloor. The technique could enable efficient monitoring of the reproductive success of the squid population, leading
one day to improved management decisions about when and where squid should be
taken by humans for food.

Market squid mating and laying eggs

Market squid inhabit waters both deep offshore and shallower nearshore in central and southern California. They are taken commercially from the same areas where they spawn, and because their life cycles are short, less than one year, the success of the commercial catch is closely related to the spawning success of the animals. Not only humans feast on the squid; it is a prey species for many fish, birds, and marine mammals. In turn, squid prey on many other organisms. Gaining a better understanding of the market squid's life cycle and the impacts of the squid fishery is important to ensure calamari remains on restaurant menus. Although the fishery seems to be sustainable off California, concerns about the large scale of the worldwide squid fishery and the lack of controls on the fishery in international waters have led the Monterey Bay Aquarium's Seafood Watch to rate squid, or calamari, as a "good alternative," rather than the "best choice" for eating.

Research efforts in the Monterey Bay National Marine Sanctuary's waters have also focused on populations of krill and their relationship to the episodes of warm and cold ocean waters known as El Niño and La Niña. Like market squid, krill are marine animals that play an important role in the ocean food chain. Although only an inch or two long, shrimp-like krill provide the main food source for the world's largest animal, the blue whale. Krill are also a food source for commercially important marine species, including market squid, salmon, and sardines. Studies by scientists from the Institute of Marine Sciences at the University of California at Santa Cruz revealed that the availability of krill to predators within Monterey Bay National Marine Sanctuary was disrupted significantly during recent El Niño and La Niña episodes. Due to the significance of krill to ocean ecosystems, Marine Sanctuary management took action, beginning in 2003, to extend California's existing ban on the commercial take of krill in state waters to the entire Marine Sanctuary and beyond. The Pacific Fishery Management Council has signaled its intention to ban the take of krill in all federal waters off the U.S. West Coast.

Because many of its resources are not easily viewed, the Monterey Bay National Marine Sanctuary places a high value on public education. In this effort the Sanctuary works with many partners, such as the nonprofit Monterey Bay Sanctuary Foundation, which has produced a series of books, posters, and maps about the Marine Sanctuary. In addition, the Sanctuary worked in partnership with the Monterey Bay Aquarium to publish *A Natural History of the Monterey Bay National Marine Sanctuary*, a richly illustrated guide. This and other publications are available at the Monterey Bay Aquarium and local booksellers.

Mapping the geology, habitat, and biodiversity of the ocean's floor is the subject of research efforts conducted within the Monterey Bay National Marine Sanctuary. Unlike the mapping and aerial photos of America's cities and towns, widely distributed over the Internet, the mapping of offshore areas has proceeded much more slowly. Interactive maps of beaches, offshore kelp forests, and submarine canyons are now available on the Internet at the website maintained by the Sanctuary Integrated Monitoring Network; for more information, see: www.mbnms-simon.org.

On the Sanctuary's website, one can check the weather forecast for coastal and marine areas, wind and wave forecasts, and satellite photos. The Sanctuary offers a wealth of written material for visitors, such as brochures with information about recreational activities and where to pursue them. Visitor centers at coastal state parks and other attractions, including the Monterey Bay Aquarium, stock books, maps, and posters about the Sanctuary. The Coastal Discovery Center, jointly operated by the Monterey Bay National Marine Sanctuary and the California Department of Parks and Recreation, opened at William R. Hearst Memorial State Beach in San Simeon in 2006, and a larger, more comprehensive Sanctuary visitor center is scheduled to open in Santa Cruz in 2008. For more information, call: 831-647-4201, or see: http://montereybay.noaa.gov.

Gray whale skim-feeding on krill

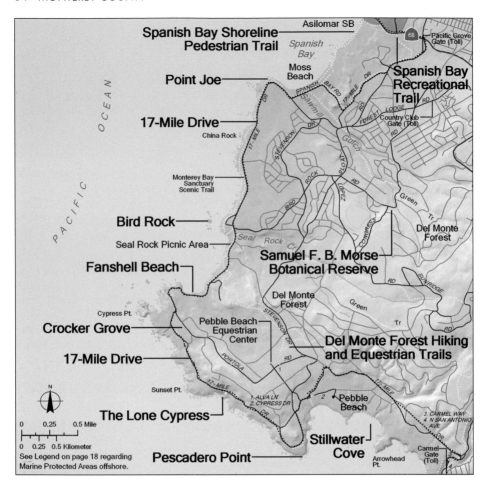

Asilomar SB
Spanish Bay Shoreline Pedestrian Trail
Spanish Bay
68
Pacific Grove Gate (Toll)
Point Joe
Moss Beach
Spanish Bay Recreational Trail
SPANISH BAY RD
17-MILE DR
DR
LODGE
FOREST
RD
RD
17-Mile Drive
China Rock
17 MILE DR
STEVENSON
Sawmill
DR
Gulch
Country Club Gate (Toll)
Monterey Bay Sanctuary Scenic Trail
ROCK
ISLAY
LOPEZ
RD
Green Tr
Del Monte Forest
Bird Rock
BIRD
CONGRESS
Seal Rock Cr
Seal Rock Picnic Area
Samuel F. B. Morse Botanical Reserve
Fanshell Beach
Del Monte Forest
Green
RD
SUNRIDGE
Cypress Pt.
Pebble Beach Equestrian Center
STEVENSON DR
Tr
RD
Del Monte Forest Hiking and Equestrian Trails
Crocker Grove
PORTOLA
RD
17-Mile Drive
17 MILE DR
1. ALVA LN
2. CYPRESS DR
2
Pebble Beach
17-MILE DR
Sunset Pt.
N
The Lone Cypress
3. CARMEL WAY
4. N SAN ANTONIO AVE
0   0.25   0.5 Mile
0   0.25   0.5 Kilometer
See Legend on page 18 regarding Marine Protected Areas offshore.
Pescadero Point
Stillwater Cove
Arrowhead Pt.
Carmel Gate (Toll)
OCEAN
PACIFIC

Fanshell Beach

17-Mile Drive

	Sandy Beach	Rocky Shore	Trail	Visitor Center	Campground	Wildlife Viewing	Fishing or Boating	Facilities for Disabled	Food and Drink	Restrooms	Parking	Fee
Spanish Bay Recreational Trail			•									
Spanish Bay Shoreline Pedestrian Trail	•	•	•									
Point Joe		•	•			•					•	
17-Mile Drive	•	•	•			•	•		•	•	•	
Bird Rock		•				•	•		•	•		
Fanshell Beach	•					•					•	
Crocker Grove											•	
The Lone Cypress		•									•	
Pescadero Point		•									•	
Stillwater Cove	•						•				•	
Del Monte Forest Hiking and Equestrian Trails			•									
Samuel F. B. Morse Botanical Reserve			•							•		

SPANISH BAY RECREATIONAL TRAIL: *Sunset Dr. and Asilomar Ave., Pacific Grove.* A pedestrian and bicycle path starts at Sunset Dr. and Asilomar Ave. and extends into the Del Monte Forest, connecting with 17-Mile Dr. The trail passes through native Monterey pine forest inland of the Inn at Spanish Bay; picnic tables are located along the way. Where the trail merges with 17-Mile Dr., bicyclists may continue along the road, a designated bicycle route. No entrance fee to the 17-Mile Dr. for pedestrians and bicyclists.

SPANISH BAY SHORELINE PEDESTRIAN TRAIL: *Sunset Dr., W. of Asilomar Ave., Pacific Grove.* A pedestrian boardwalk runs seaward of the Spanish Bay golf course, starting at Asilomar State Beach; shoulder parking. From the boardwalk, short paths provide access to the broad, fine-grained sandy beach known as Moss Beach. To protect native dune vegetation, stay on the boardwalk. At the Spanish Bay picnic area on 17-Mile Dr., the boardwalk turns into a packed path and continues south along the shoreline as far as the Seal Rock picnic area. The path offers many scenic vistas of pocket beaches and offshore rocks. At the Spanish Bay picnic area and at Seal Rock Creek, the trail connects with

the Del Monte Forest Hiking and Equestrian Trail network, which continues inland.

POINT JOE: *S.W. of Moss Beach, 17-Mile Dr.* Point Joe marks the south end of Spanish Bay. The site of several shipwrecks, the area directly offshore here experiences unusual wave patterns and turbulence resulting from ocean bottom topography and the meeting of ocean swells and currents. Two parking areas with viewing areas; a packed-earth path parallels the shore. Brandt's cormorants and other seabirds roost on nearby offshore rocks.

17-MILE DRIVE: *Between Pacific Grove and Carmel, Del Monte Forest.* This famous scenic drive traverses the edge of the Monterey Peninsula, past well-known attractions. The privately owned 17-Mile Dr. and Del Monte Forest have five entrance gates where a toll is charged for vehicles; pedestrians and bicyclists enter free. Overlooks, picnic areas, and trails along the 17-Mile Dr. are available for public use, and several golf courses and the Pebble Beach Equestrian Center are open to the public for a fee. Call: 831-647-7500.

The privately held Pacific Improvement Company opened a road in 1881 for use by guests at Monterey's Del Monte Hotel, now the Naval Postgraduate School. Horse-drawn coaches made a 17-mile round trip from Monterey along the coast of Pacific Grove, south to the Carmel Mission, and back over Carmel Hill to Monterey. The wooded area between Pacific Grove and Carmel became known as the Del Monte Forest. A stop on the tour was Pebble Beach, site of the famous golf course that opened to the public in 1919. Today's 17-Mile Dr. is part of the original sightseeing route.

Samuel F. B. Morse, the grandnephew of the inventor of Morse Code, acquired the Del Monte Forest in 1919. Morse subdivided the land while preserving much of the Monterey pine and Monterey cypress tree forest. Today, the Del Monte Forest is a private community managed by the Pebble Beach Company and developed with elaborate homes of many styles, golf courses, and hiking and equestrian trails.

BIRD ROCK: *S. of Point Joe, 17-Mile Dr.* Bird Rock and neighboring Seal Rock are hauling-out grounds for harbor seals and California sea lions. The nearshore rocks also provide summer roosting sites for brown pelicans and nesting areas for Brandt's cormorants. The Seal Rock picnic area, overlooking the ocean and nearshore rocks, has tables and restrooms. A wheelchair-accessible path runs along the shoreline from Seal Rock Creek north to the Spanish Bay picnic area.

FANSHELL BEACH: *Signal Hill Rd. and 17-Mile Dr.* Sea otters frequent the crescent-shaped cove offshore from this sandy beach, and harbor seals give birth to their young here in the spring. The Fanshell Beach overlook is closed during the harbor seal pupping season, from April 1 to June 1. Pink sand verbena and the endangered Menzies wallflower occur on the dunes, which provide habitat for the black legless lizard.

CROCKER GROVE: *S.E. of Cypress Point, 17-Mile Dr.* Native Monterey pine and Monterey cypress trees can be viewed from the margin of this 13-acre reserve; a small parking area is located on 17-Mile Dr. The nearby Cypress Point lookout offers a well-known, but somewhat constricted, view across Carmel Bay.

THE LONE CYPRESS: *S.E. of Cypress Point, 17-Mile Dr.* A solitary Monterey cypress tree clinging to a granite headland is a highly popular attraction for photographers and artists. Perhaps the best view is from the observation platform at the southeast end of the viewing area.

PESCADERO POINT: *1.5 mi. S.E. of Cypress Point, 17-Mile Dr.* Pescadero Point, a blufftop overlook with a rocky shore and tidepools below, marks the northern tip of Carmel Bay. Just north of the point is the Ghost Tree, a Monterey cypress named for its bleached-white twisted trunk and gnarled branches shaped over time by wind and sea spray. Author Robert Louis Stevenson described Monterey cypress trees as "ghosts fleeing before the wind." No fishing or picnicking allowed at the viewpoint.

Pebble Beach ca. 1910

Bird Rock

STILLWATER COVE: *End of Cypress Dr., off 17-Mile Dr., Del Monte Forest.* A small sandy beach and pier are located near the Lodge at Pebble Beach. Stillwater Cove Beach provides diving access to Stillwater Cove, which is protected from northwest swells by Pescadero Point. The rocky substrate of the cove supports giant kelp, puffball sponges, sunflower stars, plume worms, and sea hares. Harbor seals haul out on offshore Pescadero Rocks.

The beach at Stillwater Cove is open during daylight hours. Entrance to the beach is through the parking lot of the private Beach and Tennis Club. Public parking for beach access is available by reservation only; 16 parking spaces are set aside for beach users at no charge. Divers or visitors with hand-launched boats may drive into the designated loading area in the Beach and Tennis Club to drop off or pick up equipment. The parking lot is closed daily between 11 AM and 2 PM and at certain other times. For information and parking reservations, which may be made up to two weeks in advance, call: 831-625-8536.

DEL MONTE FOREST HIKING AND EQUESTRIAN TRAILS: *Off 17-Mile Dr., Del Monte Forest.* A 25-mile network of riding and hiking trails leads through the forest. The color-coded trails are marked by wooden posts or small "blazers" mounted on trees; the longest trail, marked with green, loops from the Spanish Bay picnic area uphill to the Samuel F. B. Morse Botanical Reserve and back to the shoreline. Ask at the Pebble Beach Equestrian Center, Portola Rd. near Stevenson Dr., for a trail map and information on guided horseback rides and other services. Open from 8 AM to 5 PM; call: 831-624-2756.

SAMUEL F. B. MORSE BOTANICAL RESERVE: *Bird Rock Rd. and Congress Rd., Del Monte Forest.* The forest on Huckleberry Hill includes native Monterey pine trees, growing here in an unusual association with bishop pines, and the endangered Gowen cypress, which occurs naturally only at this location and at Point Lobos State Reserve. Huckleberry Hill is an ancient marine terrace, some 800 feet above sea level, containing poorly drained claypan soils that support a stunted community, or "pygmy forest," of bishop pine and Gowen cypress. The reserve is maintained primarily to protect endangered native plants, but several hiking trails start at Congress Rd. near Bird Rock Rd. and lead through the two parcels of land that make up the reserve. The forest is a hazardous fire area; no smoking or campfires of any kind are permitted. Managed by the Del Monte Forest Foundation; call: 831-373-1293.

Monterey Pine Forest

MONTEREY PINES have been planted widely in many parts of the state and around the world. The significance of the native Monterey pine forest on California's Central Coast is somewhat obscured by the everyday appearance of the trees. Furthermore, the natural beauty of the Monterey pine forest in its grand coastal setting tends to overshadow its importance as a "gene bank" and a key component of a unique web of plant and animal life. On the Monterey Peninsula, the Monterey pine forest is associated with at least 19 plant species and 17 wildlife species that are designated as having special status, due to their rarity or significance.

The Monterey pine forest grows on the Monterey Peninsula, which enjoys a cool, maritime climate influenced by onshore breezes and the upwelling of cold ocean waters from arms of the deep Monterey Submarine Canyon that hug the peninsula. The area's modest rainfall of 15 to 20 inches per year occurs mostly in the winter months, but summertime fog drip provides significant additional precipitation. Monterey pine trees, and the Monterey cypress trees that also grow on the Monterey Peninsula, once occupied a much wider territory. When the California coast became hotter and drier between 8,000 and 4,000 years ago, Monterey pines persisted in three locations: the Monterey Peninsula, Año Nuevo Point near the San Mateo/Santa Cruz County line, and Cambria in San Luis Obispo County. Monterey pines are also found on two islands off the coast of Baja California.

The Monterey pine is a closed-cone conifer, meaning that its seeds are held closely on the cones, sometimes for years, until released by very hot weather or a fire. Following a light ground fire, pine seedlings sprout readily after being watered by winter rains. As development of homes and golf courses in the Monterey pine forest has taken place, fires have been suppressed, and reproduction of the native trees has suffered.

As restricted as it is, the Monterey pine forest found on the Monterey Peninsula has at least 11 distinct subtypes differentiated in part by soil type, including marine terrace deposits, dunes, alluvial deposits, and soils over shale and granite bedrock. On the Monterey Peninsula there are six marine terraces, and Monterey pines are found on all of them. The first terrace, the one closest to the shoreline, is the youngest. Farther from the shoreline, the terraces are progressively older and higher. This configuration of landforms is known as an "ecological staircase." Sand dunes, also of varying age, have accumulated on parts of the first, second, third, and fourth marine terraces, and Monterey pine trees occur on parts of these dunes. On the fifth marine terrace at an elevation of 320 to 540 feet, Monterey pines are somewhat stunted in size and mixed with coast live oak and stands of bishop pine forest. The Monterey pine trees grow to their fullest size locally, 80 to 100 feet tall, on the shale and granitic bedrock soils. Some researchers think that the forest subtypes are disjunctive and correspond to the marine terraces, whereas others see a simple gradient in forest characteristics. In any event, the Monterey pine forest is remarkably variable in growth characteristics and understory associations.

Monterey pine, called radiata pine or New Zealand pine in the timber industry, is a major commercial species in New Zealand, Australia, Chile, Spain, Kenya, South Africa, and elsewhere. There are nearly 10 million acres of commercial stands of the pine, which is the world's most-planted conifer. Monterey pine trees planted in New Zealand reach close to 200 feet tall. Artificially selected trees attain heights of over 100 feet in little more than 20 years, and trees reach maturity in less than 30 years. Given ample moisture and the warm temperatures of the tropics or subtropics, Monterey pines grow year-round and exhibit wood that lacks annual rings. The commercial appeal of the species comes from its rapid growth, large size when growing conditions are ideal, and high-quality lumber. Monterey pine plantations are a mixed blessing in the countries where they are grown. On the one hand, they probably have prevented the lumbering of many thousands of acres of native forests; on the other hand, they are exotic invaders of those same forests.

The world's commercial forests of Monterey pine trees differ from the few relict, native stands in California and Baja California not only in their appearance, but also in the fact that the farmed stands are restricted genetically, having been propagated selectively for fast, tall growth. The tree farms thus lack the genetic variability of the native Monterey pine forest. The preservation of a bank of different strains of Monterey pines is all the more important because of the challenge presented by pine pitch canker, a fungal disease that attacks Monterey pines and for which there is no available treatment. Since 1992, the forest on the Monterey Peninsula has been attacked by pine pitch canker. Some trees appear to be tolerant of the disease, however, suggesting that the genetic diversity of the native Monterey pine trees may one day help provide solutions for this and other diseases in Monterey and around the world.

Monterey pine forest community, Samuel F. B. Morse Botanical Reserve

Monterey pine

The **Monterey pine** (*Pinus radiata*) is closely related to bishop pine and knobcone pine, however it is distinguished from the former by having needles in groups of threes, and from both by having blunt scales on the cone. All three of these pine species are known as closed-cone pines. The cones of Monterey pine remain closed until opened by the heat of a forest fire; the abundant seeds are then discharged and begin a new forest. The cones may also burst open, with a snapping sound, in hot weather. The Monterey pine can be seen at many locations on the Monterey Peninsula. Point Lobos State Reserve is a great place to hike in a Monterey pine forest.

Monterey cypress

The **Monterey cypress** (*Cupressus macrocarpa*) occurs naturally in only two groves near Monterey. These native groves are protected within the Point Lobos State Reserve and the Del Monte Forest at Cypress Point. The trees grow on the exposed rocky headlands along the ocean. If you take a walk along the coast at Point Lobos you will see that the cypress trees, subjected to the constant sea breeze and salt spray, grow in a unique twisted and contorted fashion. The Costanoan Indians were known to make a decoction of Monterey cypress foliage to treat rheumatism. Although the Monterey cypress is widely planted as a landscape tree, the California Native Plant Society considers this a rare species, based on its limited natural occurrence.

Yadon's rein orchid

Yadon's rein orchid (*Piperia yadonii*), also known as Yadon's piperia, was first collected in 1925 in open pine forest near the town of Pacific Grove. At that time, it was identified as *Piperia unalascensis*. In a recent taxonomic analysis of the genus, two new species were identified. One of these, *P. yadonii*, was named after Vern Yadon, former director of the Museum of Natural History in Pacific Grove. As with other orchids, germination of Yadon's rein orchid seeds likely involves a symbiotic relationship with a fungus. One to several years after germination, orchid seedlings produce their first basal leaves. Yadon's rein orchid is found within Monterey pine forest and maritime chaparral communities in northern coastal Monterey County. This lovely orchid is listed as endangered under the federal Endangered Species Act.

Like most owls, the **great horned owl** (*Bubo virginianus*) is active at night and is almost silent in flight. This majestic and stately bird has large tufts of feathers on its head that look like horns. These feathers are sometimes referred to as "ear tufts," but they have nothing at all to do with hearing. Great horned owls prey on rabbits, other small mammals, and birds. They often build their nests in the hollows in old trees. Listen for the great horned owl when you take a night hike in the Monterey pine forest. You are likely to hear its distinctive call, a series of muffled hoots: *hoo hoodoo hoooo hoo.*

Great horned owl

The **dusky-footed wood rat** (*Neotoma fuscipes*) is a type of "packrat." The wood rats build large cone-shaped houses, two to five feet high, out of sticks and leaves. A house or den is usually located at the base of a tree, but sometimes the wood rats locate the house up in the tree's branches. The wood rat builds a nest inside the house, using shredded grass, leaves, and other materials. A single den is often the result of work by several generations of wood rats. The dusky-footed wood rat caches food such as leaves, berries, and nuts in chambers inside its nest. Sometimes the cache chambers may provide a home for other small mammals, frogs, and invertebrates. Point Lobos State Reserve is a good place to look for wood rat nests.

Dusky-footed wood rat

The **gray fox** (*Urocyon cinereoargenteus*) resembles a small dog with a big, bushy tail. It has a silvery gray mantle, conspicuous reddish or buffy patches on the throat and belly, and a black tip on the end of the tail. It is generally nocturnal and very shy, but occasionally it may be seen during the day hunting for food or "mousing." The gray fox feeds on small rodents, birds, berries, insects, and fungi. Gray foxes are unique among canids in their ability to climb trees. They have strong, hooked claws that allow them to get sufficient grip to scramble up trees. This unusual skill allows them to escape enemies, eat fruits or other foods found in trees, or just lounge in the comfort of the tree's canopy.

Gray fox

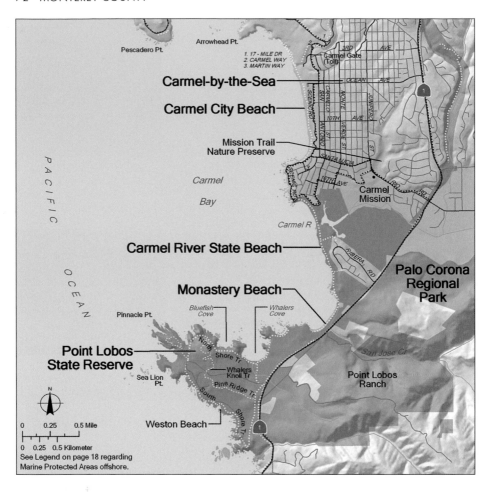

Carmel City Beach

Carmel Area

	Sandy Beach	Rocky Shore	Trail	Visitor Center	Campground	Wildlife Viewing	Fishing or Boating	Facilities for Disabled	Food and Drink	Restrooms	Parking	Fee
Carmel-by-the-Sea	•	•	•				•		•	•	•	
Carmel City Beach	•						•		•	•		
Carmel River State Beach	•	•	•			•	•		•	•		
Monastery Beach	•	•					•		•	•		
Point Lobos State Reserve	•	•	•	•		•	•		•	•		•
Palo Corona Regional Park						•						

CARMEL-BY-THE-SEA: *Off Ocean Ave., W. of Hwy. One, 2 mi. S. of City of Monterey.* An utterly charming community in a forest. In Carmel, the post office serves as a town meeting-place; home mail delivery is not possible because houses bear name plaques rather than street numbers. The Great Sand Castle Contest held annually in October is serious business; the event is sponsored in part by the local chapter of the American Institute of Architects. The long-running summertime Carmel Bach Festival presents concerts at the Carmel Mission church and elsewhere; call: 831-624-2046.

CARMEL CITY BEACH: *Foot of Ocean Ave., Carmel.* The spectacular, fine white sand beach, composed of quartz and feldspar from eroded granite, is popular for picnicking and strolling. Parking is at the end of Ocean Ave., but can be very tight on weekends. Along Scenic Rd., which hugs the coast, a blufftop trail extends south from 8th Ave. to Martin Way. Street parking; eight stairways and a ramp provide access to the beach. Volleyball court on back beach. Dogs allowed on the beach under voice control; pick up after your pet. Beach fires permitted only south of 10th Ave. Call: 831-624-3543.

Carmel-by-the-Sea

Non-wheelchair-accessible restrooms are at the end of Ocean Ave.; wheelchair-accessible portable toilets located on Scenic Rd. at Santa Lucia Ave.

CARMEL RIVER STATE BEACH: *Scenic Rd. at Carmelo St., Carmel.* A splendid mile-long crescent of fine white sand lies at the mouth of the Carmel River. The beach is bounded by rock outcroppings to the north and the jagged profile of Point Lobos to the south, with forested slopes rising in the distance. Beach access, from the parking lot on Scenic Rd., has been complicated in recent years by the natural meandering of the Carmel River. When the river mouth moves north, there is sometimes only a narrow strip of sand connecting the parking lot to the beach.

The lagoon and wetlands near the mouth of the river are a bird sanctuary. In summer, great flocks of California brown pelicans rest on the sandbar, cruising low over the water to the nearby ocean. On a still afternoon, the enormous wings of the birds sound like sheets of canvas flapping in the breeze. Gulls and shorebirds rest and feed along the lagoon, and steelhead trout spawn

upstream in the river. Divers use the beach for access to the nearby underwater kelp forest, where the rocky bottom supports sea lemons and strawberry anemones. Ocean swimming and wading are dangerous here. Restrooms with running water are located next to the parking lot. Another access point to Carmel River State Beach is located south of the Carmel River, off Hwy. One at the end of Ribera Rd.

The Carmel River forms the northern boundary of the California Sea Otter Game Refuge, which extends south to San Luis Obispo County. The Big Sur coast is one of only two remaining areas in which sea otters are found; the other is in Alaska. The best place to see the otters close up is at the Monterey Bay Aquarium; the animals can also be glimpsed sometimes in the waters surrounding the Monterey Peninsula or at Point Lobos.

MONASTERY BEACH: *Off Hwy. One, 1.5 mi. S. of Rio Rd. intersection, Carmel.* Monastery Beach, part of Carmel River State Beach, offers fine views of Carmel and Point Lobos and is a popular diving beach. Upwelling

Carmel River State Beach

Carmel River

of nutrients from the nearby Carmel Submarine Canyon supports a rich marine life; a marine protected area designation by the California Department of Fish and Game provides special protection to marine resources in Carmel Bay from Pescadero Point to Point Lobos. The deep canyon also contributes to extremely dangerous surf conditions at this steep, coarse-grained beach, which is unsafe for swimming. The best diving access is from the north end of the beach near rocks and the adjoining kelp forest. Restrooms and roadside parking available; call: 831-649-2836.

On the inland side of Hwy. One, opposite Monastery Beach, the California Department of Parks and Recreation plans to acquire the Point Lobos Ranch property, which includes stands of native Monterey pine forest and maritime chaparral. Call: 831-649-2836.

POINT LOBOS STATE RESERVE: *Off Hwy. One, 2.2 mi. S. of Rio Rd. intersection, Carmel.* The significance of the natural resources on this forested rocky headland is fully matched by the grandeur of the setting. Monterey cypress and Gowen cypress trees grow naturally only in the Point Lobos State Reserve and in the Del Monte Forest; the reserve also contains one of the world's few native stands of Monterey pine. Wrentits and white-crowned sparrows nest in coastal scrub vegetation, and yellow-rumped warblers winter here. Bobcats, black-tailed deer, and dusky-footed wood rats also inhabit the reserve. Look for wintering monarch butterflies along the Whalers Knoll Trail. Offshore, sea otters raft in large stands of giant kelp, which shelter rockfish, cabezon, and lingcod; Pacific loons may be seen in winter.

Point Lobos State Reserve is a unit of the state park system and is maintained as a pristine area where all plants, rocks, wood, seashells, and animal life are protected and may not be disturbed, removed, or collected. Trails and picnic areas are available; docents and park rangers lead walks and answer questions. The reserve opens daily at 9 AM; no admittance after 6:30 PM in summer and 4:30 PM in winter. When full, the reserve allows vehicle entry only on a one-out, one-in basis; no trailers allowed, and vehicles over 20 feet in length are not allowed during busy periods. No pets; fee for entry. Call: 831-624-4909.

Black-tailed deer, Point Lobos State Reserve

Diving and snorkeling is permitted only in Whalers and Bluefish Coves and is limited to 15 pairs of divers per day by permit only, available at the entrance gate. Advance diving or kayak reservations recommended, and for weekends, essential; call: 831-624-8413. Within the Point Lobos State Marine Reserve surrounding Point Lobos, no fishing is allowed and no aquatic plants or animals may be taken or disturbed.

PALO CORONA REGIONAL PARK: *E. of Hwy. One, S. of Carmel River Bridge.* This new 4,500-acre park is scheduled to open during 2006 under an interim arrangement, while long-range plans for visitor use and resource protection are prepared. Managed by the Monterey Peninsula Regional Park District; for information, call: 831-372-3196. Public acquisition of the Palo Corona property was supported by the Wildlife Conservation Board, the State Coastal Conservancy, the California Department of Parks and Recreation, and the Monterey Peninsula Regional Park District. Funding came from Proposition 40, approved by California voters in 2002; the Big Sur Land Trust; and the Nature Conservancy. The new parklands will link Carmel River State Beach and Garrapata State Park with the Los Padres National Forest, creating a ten-mile-long area for public recreation and habitat protection. The property includes stands of native Monterey pine trees and old-growth redwood forest, and habitat for the condor, black bear, mountain lion, tiger salamander, Smith's blue butterfly, and steelhead trout.

Point Lobos State Reserve

The splendor of the rocky coast of Point Lobos owes much to its geology. Two rock types make up the Point Lobos peninsula. One is an igneous unit, a granodiorite, and the other is a sedimentary unit, the Carmelo Formation, that is made up of sandstone, conglomerate, and shale.

Granodiorite is an intrusive igneous rock similar to granite; that is, it cooled from a molten state several miles below the earth's surface. Because it was insulated by the overlying rocks, it cooled slowly over several million years and formed large, interlocking crystals of feldspar, quartz, and lesser amounts of dark-colored minerals. This particular body of granodiorite has been dated at 100–110 million years old.

A large gap in time separates the cooling of the granodiorite and the deposition of the overlying Carmelo Formation, which is about 60 million years old. The younger rocks were not simply deposited horizontally on top of the granodiorite, but against it in what is known as a "buttress unconformity." In three dimensions, a canyon can be seen to be carved into the granodiorite. The Carmelo Formation fills this canyon.

At Sea Lion Point the Carmelo Formation consists of alternating conglomerate and sandstone layers. These coarse rocks indicate that they were deposited by strong water currents. Very sparse fossils found in these units indicate that they are marine deposits.

At Weston Beach the Carmelo Formation consists of alternating beds of fine sandstone and shale known as "turbidites," the deposit left by a "turbidity flow." Turbidity flows represent the sudden release of a sediment mass, which quickly incorporates large amounts of water and moves downslope as a density-driven, fluid mass. When the mass loses energy at the base of a slope, it leaves a characteristic sequence reflecting the decelerating flow.

The buttress unconformity and outcrop pattern of the Carmelo Formation strongly suggest that the unit represents the fill of a submarine canyon cut into the basement granodiorite at Point Lobos. The coarse conglomerates exposed at Sea Lion Point record the highest current velocities encountered within the formation. This suggests that the axis of the main channel of the canyon must have crossed near here. As one moves southeast toward Weston Beach, the grain size decreases, suggesting that one is moving away from the main channel.

The Point Lobos peninsula preserves the convergence of two submarine canyons. In fact, the canyon system is remarkably similar in scale and in orientation to the modern Monterey–Carmel Submarine Canyon system, which lies just offshore. By looking at the submarine canyon deposits preserved on Point Lobos, visitors get a dry look at conditions on the floor of the Monterey Submarine Canyon.

Interbedded sandstone and shale of the Carmelo Formation

Point Lobos

Rocky Shore

B ETWEEN the high and low tide marks lies a strip of shoreline that is regularly covered and uncovered by the advance and retreat of the tides. This meeting ground between land and sea is called the intertidal. The plants and animals inhabiting this region are hardy and adaptable, able to withstand periodic exposure to air and the force of the pounding surf. Intertidal communities occur on sandy beaches, in bays and estuaries, and on wharf pilings, but the communities of rocky shorelines are perhaps the most diverse and densely populated. Rock faces, crevices, undersides of rocks, and tidepools each support an array of species.

The plants and animals of the intertidal are subject to a range of conditions not encountered in the relative stability of the deep ocean. Three factors—substrate, wave shock, and exposure to drying—are important in determining the types of organisms found in a given intertidal community. Sessile, or attached, organisms are typical of rocky shores, in contrast to sandy or mud substrates, which support an abundance of burrowing animals. Wave action on rocky headlands and exposed outer coasts can be tremendous, and only the most tenacious organisms survive. The most important factor in determining where marine organisms occur in the intertidal, however, is the ability to withstand desiccation and overheating while exposed to air by low tides.

The extent to which an organism is exposed to air is largely determined by its vertical position in the intertidal region and the pattern of the tides. On the California coast, there are two daily high tides and two low tides, of varying magnitude, so that each day there is a higher high tide, followed by a lower low tide, a lower high tide, and a higher low tide. The mean high tide level is the average of all daily high tides.

The intertidal can be divided into vertical zones based on the frequency and duration of exposure to water or air. A model proposed by Ed Ricketts and Jack Calvin in *Between Pacific Tides* divides the rocky shore into four zones characterized by different species of algae and animals. The location of these zones varies with exposure; strong wave action tends to widen the zones, whereas in sheltered areas of quiet water, the zones are narrow.

The uppermost horizon is the area above the mean high tide level and includes the splash zone that only receives spray from waves. Almost always exposed to air, the splash zone is covered with a thin sheet of black photosynthetic bacteria—"blue-green algae"—and the species found here are quite specialized. Below the splash zone is the second zone, the high intertidal, extending to the mean higher low tide level. Less harsh than the splash zone, this zone is home to a larger diversity and number of plants and animals. Large algae occur here, providing shelter for small animals and a food source for herbivores.

The middle intertidal, the third zone, extends to the mean level of the lower low tide— the zero tidal datum of the tide tables (average sea level is about +3 feet relative to this reference level). This zone is covered and uncovered by the tides twice a day and it teems with life, including mussels, barnacles, and anemones. The low intertidal, by far the richest and most densely populated zone, is the lowest of the four zones, extending from tidal datum to the level of the lowest low tide. This zone is uncovered only at "minus" tides, and it is habitat for sponges, nudibranchs, sea stars, and sea urchins. Many species would thrive lower in the intertidal than they are normally found, but are confined to higher levels by competitors, predators, or herbivores. For example, acorn barnacles are capable of living over a wide vertical range, but in the middle intertidal, the predatory rock snail eats juvenile barnacles that settle there. In the high intertidal, where the rock snail cannot survive, the barnacles flourish.

Acorn barnacle

Acorn barnacles (*Balanus* sp.) are abundant in the high intertidal zone of exposed and semi-exposed rocky shores. Acorn barnacles are crustaceans, closely related to shrimp. However, barnacles bear little similarity to their shrimp cousins. Each barnacle builds its own cone-shaped fortress, cemented to the rocky substrate. From within its shell, the acorn barnacle extends feathery "feet" with which it collects plankton and detritus from the churning water. When the tide goes out, the barnacle closes the opening of its shell to conserve moisture.

Opalescent nudibranch

During low tides, it is possible to find **opalescent nudibranchs** (*Hermissenda crassicornis*) in the kelp, generally in the low intertidal zone. The opalescent nudibranch is one of the most abundant nudibranchs, or "sea slugs," found on the Pacific Coast. Opalescent nudibranchs feed on hydroids and other invertebrates. Their bright coloration serves as a warning to predators that the "cerata," or finger-like projections on their backs, are tipped with small, stinging structures. Like most other sea slugs, they are hermaphroditic. Therefore, reproduction can occur between any two individuals.

Monkeyface eel

Monkeyface eels (*Cebidichthys violaceus*) are not true eels, but rather are members of the prickleback family. Their distorted faces are a bit grotesque and give them an eel-like appearance. They act like eels as well, slithering into crevices under rocks and seaweed. This long slender fish, about one to two feet long, has the unfishlike ability to breathe and survive out of water temporarily. Typical habitat for monkeyface eels includes rocky areas with ample crevices, such as high and low intertidal tidepools, jetties and breakwaters, and relatively shallow subtidal areas, particularly kelp beds.

The **sea lemon** (*Archidoris montereyensis*) is a large, flattened sea slug, up to seven inches long, with two horn-like projections at one end and a tuft of gills at the other. The mantle is sprinkled with black dots. The sea lemon primarily feeds on encrusted sponges, especially *Halichondria* and *Haliclona* spp. It uses its radula (a flexible, tongue-like organ having rows of tiny teeth on the surface) to scrape the sponge off the substrate and then ingest it. Unlike other sea slugs, the sea lemon cannot store the toxins of its prey. However, with its bright coloration, it may be mimicking other sea slugs that have that ability and are therefore poisonous to predators. The sea lemon can be found in subtidal, shaded areas where sponges grow.

Sea lemon

The adult **Heermann's gull** (*Larus heermanni*) is one of the most beautiful and easily recognized North American gulls. With its dark gray upper parts, white head with a bright red, black-tipped bill, it is like no other species. Heermann's gulls are almost exclusively coastal, ranging only a few dozen miles out to sea to forage. They feed on small pelagic fish such as sardines and anchovies, and occasionally on crustaceans, mollusks, and other marine organisms scavenged from the beach and kelp beds. This medium-sized gull species can be quite aggressive, and is often seen harassing other birds to make them drop food items. Heermann's gulls nest primarily on islands in the Gulf of California.

Heermann's gull

The **ruddy turnstone** (*Arenaria interpres*) is a short-legged shorebird that is about seven inches in length. It has a short, dark bill that is slightly upturned at the end and striking, black-and-white markings on its head. In winter, it can be found in a variety of habitats along the coast including mudflats, beaches, and the rocky intertidal. However, these small birds nest solely in the high arctic tundra. The turnstone gets its name from its habit of turning over stones when it looks for food. It is also known as the seaweed bird, because it often feeds among the kelp at low tide.

Ruddy turnstone

Point Lobos SR

Point Lobos
Ranch

Carmel
Highlands

Yankee Pt.

Palo Corona
Regional
Park

Garrapata State Park

Soberanes
Canyon Tr

Gates 7 & 8

Soberanes Cr

Lobos Rocks

Soberanes Point

Gate 10
Gate 11

1

Granite Cr

PACIFIC

Doud Cr

Gate 17

Gates 18 & 19

Garrapata Beach

Garrapata Cr

Joshua Creek
Ecological Reserve

Viewpoint

Kasler Pt.

Rocky Pt.

GARRAPATOS RD

PALO COLORADO RD

**Mill Creek
Redwood
Preserve**

Turner Creek Tr

OCEAN

Rocky Creek Bridge

Bixby Creek Bridge

Bixby Cr

Los
Padres
NF

Mill Cr

N

Hurricane Point

OLD COAST RD

Bottcher's Gap

Skinner Ridge Tr

Sierra Cr

R

0   1   2 Miles

0   1   2 Kilometers

Brazil Ranch

Little Sur R

Little   Sur

See Legend on page 18 regarding
Marine Protected Areas offshore.

Rocky Ridge Tr

San

Jose Cr

View from Hurricane Point

Big Sur North

	Sandy Beach	Rocky Shore	Trail	Visitor Center	Campground	Wildlife Viewing	Fishing or Boating	Facilities for Disabled	Food and Drink	Restrooms	Parking	Fee
Garrapata State Park	•	•	•			•				•	•	
Soberanes Point		•	•			•				•		
Garrapata Beach	•	•	•							•		
Viewpoint		•								•		
Bixby Creek Bridge										•		
Hurricane Point		•								•		
Brazil Ranch		•										
Mill Creek Redwood Preserve						•						
Bottcher's Gap		•		•						•	•	•

GARRAPATA STATE PARK: *Hwy. One, 3.3 mi. S. of Point Lobos State Reserve.* A little advance planning will make a visit to Garrapata State Park more rewarding. This day-use-only park stretches four miles along the seaward side of Hwy. One and more than two miles inland, but there are few facilities, and park entrances are limited to easily missed numbered roadside gates, next to pull-outs. Hwy. One traffic moves fast here; use caution when slowing to park along the highway. Look carefully for the trails, which are sometimes overgrown. Poison oak grows magnificently here, as do other native plants typical of the coastal bluffs.

At the north end of Garrapata State Park, four southbound pull-outs near the 67.0 milepost marker offer access to a steep bluff vegetated with spring wildflowers and sticky monkeyflower. There are good views of rocky shore but no safe access to the water. Near milepost 66.0 is roadside gate #7 and a steep path overlooking Soberanes Creek, where it cascades to a cobble beach. On the inland side of Hwy. One opposite gate #7, the Rocky Ridge Trail climbs five miles to the 1,977-foot summit of Doud Peak; bicycles are allowed on the trail. A couple of hundred yards south of gate #7 on Hwy. One, a hedgerow of Monterey cypress trees and pull-outs on both sides of the highway mark gate #8, leading to a blufftop with a particularly striking view of rocky shore. Opposite gate #8, the Soberanes Canyon Trail follows Soberanes Creek inland past a barn and up a long, narrow valley through a redwood grove, connecting to the Rocky Ridge Trail. Excellent views, and spring wildflowers, but the trail is steep and somewhat eroded; use caution. Pit toilet only; dogs not allowed. Call: 831-649-2836.

SOBERANES POINT: *Hwy. One, 4.9 mi. S. of Point Lobos State Reserve.* Soberanes Point is marked by steep, cone-shaped Whale Peak, rising seaward of Hwy. One. Several small pull-outs provide parking. Gate #10 opens on two paths, the southerly one extending in a two-mile-long loop around the perimeter of Soberanes Point, and the northerly path leading to the 460-foot crest of Whale Peak, with striking views of rocky shore and offshore kelp forest. Watch for a wrentit or a buckeye butterfly among the dense growth of coffeeberry, ceanothus, and poison oak; gray whales may be sighted during winter months, passing on their annual migration. Sea lions rest and brown pelicans roost on Lobos Rocks, offshore from Soberanes Point. Part of Garrapata State Park; dogs not allowed on trails.

South of Soberanes Point, a pull-out marked with a brown Coastal Access sign is next to gate #11, where a short, steep path leads to an ocean-viewing bench inscribed with the words: "Into her arms he dove to begin his new journey." Gate #17 at milepost 64.0 south of the Granite Creek Bridge opens to a short, steep path leading to another bench with a fine view of the steep coastal terrace and rocky shore below.

GARRAPATA BEACH: *Hwy. One, 7.3 mi. S. of Point Lobos State Reserve.* Garrapata Beach is one of the few publicly accessible sandy beaches between the mouth of the Carmel River and Andrew Molera State Park. Aquamarine ocean and a sweep of white sand are tempting to visitors, but the steep beach and rough surf make swimming unsafe and prohibited. Strolling, sunbathing, and solitude-seeking are popular here. Parking for the beach is in roadside pull-outs on both sides of Hwy. One; Monterey-Salinas Transit route signs mark the pull-outs. Paths lead seaward from half-hidden gates #18 and 19, on the north and south sides of Doud Creek, to the sandy beach and to a short north-south trail along the edge of the blufftop. At the south end of the beach, a stair accessible from gate #18 is currently closed due to erosion. Camping not allowed; dogs permitted on leash. No facilities; call: 831-649-2836.

VIEWPOINT: *Hwy. One, .5 mi. S. of Garrapata Beach.* A view of rocky Otter Cove is available from a parking area on the west side of Hwy. One, north of Kasler Point.

BIXBY CREEK BRIDGE: *Hwy. One, 11 mi. S. of Point Lobos State Reserve, Big Sur Coast.* The Bixby Creek Bridge is a major landmark of the Big Sur Coast. The bridge's completion in 1932 filled in one of the most challenging gaps in the construction of Hwy. One on the Big Sur coast. Before the bridge opened, Big Sur-bound travelers from Monterey headed east on the Old Coast Rd., which took an inland route to avoid the impassable canyons and steep cliffs at the mouth of Bixby Creek. The bridge deck is 260 feet high, exceeding the height above water of the Golden Gate Bridge by some 40 feet. There is a small southbound pull-out located on the north side of the bridge. The similar but smaller Rocky Creek Bridge, also with a small view pull-out on the north side, is located a half-mile north of the Bixby Creek Bridge.

HURRICANE POINT: *Hwy. One, 1.3 mi. S. of Bixby Creek Bridge, Big Sur Coast.* Dramatic views are available from pull-outs on Hwy. One, which reaches an elevation at Hurricane Point of nearly 500 feet above sea level. Tufted puffins and common murres nest on the steep cliffs.

BRAZIL RANCH: *S. of Bixby Creek Bridge, E. and W. of Hwy. One, Big Sur Coast.* This 1,200-acre historic ranch property, including coastline between the Bixby Creek Bridge

Bixby Creek Bridge

Soberanes Point

and Hurricane Point and inland redwood forest along Bixby and Sierra Creeks, has been acquired by the U.S. Forest Service for addition to Los Padres National Forest. The nonprofit Big Sur Environmental Institute manages the land and is preparing plans for public access, to include uses such as hiking, photography, and small-scale educational purposes. Call: 831-625-3564.

MILL CREEK REDWOOD PRESERVE: *Off Palo Colorado Rd., E. of Hwy. One, Big Sur.* This forested and mountainous area is planned for future public access on a limited basis by advance reservation. For information, call the Monterey Peninsula Regional Park District ranger station: 831-622-0598.

BOTTCHER'S GAP: *End of Palo Colorado Rd., 8 mi. E. of Hwy. One, Big Sur.* Bottcher's Gap is a day-use area, campground, and trailhead for hiking into the backcountry. There are 11 campsites with fire rings and barbecue grills among oak and madrone trees. Vault toilets; bring your own water. Fee for camping. Part of Los Padres National Forest; for information, call: 831-385-5434.

From Bottcher's Gap, the Skinner Ridge and Ventana Trails lead equestrians and backpackers some 21 miles into the Ventana Wilderness Area and to 4,853-foot-high Ventana Double Cone. Other trails lead to the upper watershed of the Little Sur River. South of Bottcher's Gap is Pico Blanco, a 3,709-foot mountain with a white top of metamorphosed limestone, or calcium carbonate.

Garrapata Beach

Rivers and Streams

ON THEIR WAY to the ocean, California's coastal rivers and streams flow through the canyons and valleys of coastal mountains, linking forest, chaparral, grassland, and marsh. Riparian woodlands develop along stream banks and floodplains, and coastal wetlands and estuaries form where the rivers enter the sea. Rivers transport nutrients, sediments, and oxygen through the watershed, and life flourishes in their path.

Streams and the surrounding riparian woodlands support numerous animal species, including frogs, salamanders, snakes, and river otters. Western sycamore trees, alders, and willows grow along the stream banks and attract large numbers of resident and migratory birds. An entangling understory of shrubs, flowering plants, and vines provides sites for nesting, shelter, and shade for many animals. Algae and mosses proliferate in the water and on rocks. Leaves swept into the current decompose, adding nutrients and organic matter. Insects thrive here and in turn provide an abundant food source for invertebrates, fish, and birds. Anadromous fish such as salmon and steelhead migrate from the sea to fresh water to spawn and depend on well-oxygenated streams and gravelly streambeds as spawning sites. River runoff, the amount of water discharged through surface streams, is determined by a combination of factors, including local geology, topography, drainage area, and rainfall patterns.

Limekiln Creek, Big Sur

The **California buckeye** (*Aesculus californica*) provides a perfect hideaway for summer naps, as the broad, pale green leaves provide a wonderful canopy to protect you against the heat of the sun. Each leaf has five leaflets, borne in a palmate arrangement. In May, the tree is covered with beautiful, fragrant flowering spikes of four- to five-petaled white flowers with long stamens. Although most of the flowers of this plant are pollinated, the tree aborts most of the young fruit. Its branches can only support a few of the heavy fruit. In the fall, big shiny chestnuts can be found on the ground at the base of the tree. The fruits are poisonous, and Native Americans poured an extract of the chestnuts into pools to stun and then catch fish. Buckeye trees grow next to springs and creeks in the coastal mountains.

California buckeye

The **western sycamore** (*Platanus racemosa*) is a deciduous, riparian tree, with five-fingered, star-shaped leaves. The most striking feature of this tree is the splotchy white and brown bark. Young greenish-gray bark exfoliates (peels off), leaving almost pure white inner bark; older bark is thicker, furrowed, and dark brown. This gives the sycamore a jigsaw puzzle-like appearance. Found not far from water, it can reach a height of 90 feet. The sycamore is home to the western tiger swallowtail butterfly. Offering shade from the summer sun, the tree affords protection and nutrients to the larva as the caterpillar becomes an adult.

Western sycamore

Coastal black gooseberry (*Ribes divaricatum*) is a deciduous, spiny shrub with beautiful flowers, edible fruit, fragrant foliage, and wonderful fall color. Gooseberries are loved by many birds including the hermit thrush and the American robin. Coastal black gooseberry plants have pendulous flowers that droop in bunches at the tips of branches from March to June. Each flower consists of a short tube, at the mouth of which are turned-back, greenish-red sepals and five white petals. Blackish berries covered in a gray-blue bloom replace flowers in late spring and summer. Despite the rather bland taste of the berries, Native Americans collected and ate them fresh.

Coastal black gooseberry

Belted kingfisher

The **belted kingfisher** (*Ceryle alcyon*) is unlikely to be confused with any other bird. Its huge bill, large head, shaggy crest, and coloring are distinctive. In addition, it has a characteristic call, consisting of a series of harsh, wooden, rattling notes of great carrying power. The belted kingfisher is bluish-gray above, with a white belly and a white ring around the neck. The belted kingfisher is one of only a few North American bird species in which the female is more colorful than the male. Females have a chestnut band across the belly that is absent in males. Belted kingfishers, as their name suggests, subsist mostly on fish. They typically sit at a waterside perch watching for prey, then they make steep dives head first into the water. Listen for the loud, rattle-like call of a belted kingfisher as it flies along the riparian corridor.

Green heron

The **green heron** (*Butorides virescens*) is a small, stocky, wading bird common in riparian areas. However, while it may be common, it is generally difficult to see. Like other herons, it stands motionless waiting for small fish to approach within striking range. The green heron is known to drop bait onto the surface of the water to attract small fish that it then seizes with a jab of its bill. When afraid or in flight, the green heron may give an explosive alarm call that sounds like *skeow*. The green heron is a rather wary and solitary bird. When alarmed, it erects its short, reddish crest, straightens its neck, and nervously flicks its short tail. The bird usually goes unnoticed until it flushes from the edge of the water and flies off, uttering its sharp call.

Foothill yellow-legged frog

Foothill yellow-legged frogs (*Rana boylii*) frequent shallow, slow, gravelly streams and rivers. They are rarely encountered far from permanent water. Yellow-legged frogs are not smooth in appearance like other frogs; their skin is bumpy like a toad's. Foothill yellow-legged frogs may be brown, gray, or rust red in appearance. They rely heavily on camouflage for their survival. They sun themselves on the banks of rocky creeks and blend in with their surroundings. If you approach one, you may be in for a surprise when a stone leaps into the creek and swims away. The foothill yellow-legged frog is considered a species of special concern by the California Department of Fish and Game.

The **Santa Lucia slender salamander** (*Batrachoseps luciae*) is a small, blackish-brown salamander measuring from one-quarter to nearly two inches in length from snout to vent. The salamander breathes through its smooth, moist, thin skin. It has short limbs, a narrow head, a long slender body, and a long tail, which gives this species a worm-like appearance. You may find the Santa Lucia slender salamander on rainy or wet nights when the temperature is mild. The salamander retreats underground when the soil dries or when the air temperature drops to near freezing. If you look under rocks or logs in moist areas, you may uncover one. It is thought that the Santa Lucia slender salamander may spend its entire lifespan in no more than a few square yards. This will not surprise you when you see how tiny its legs are.

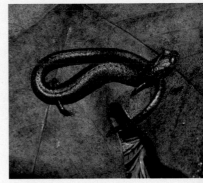

Santa Lucia slender salamander

The **western tiger swallowtail** (*Papilio rutulus*) is one of our largest butterflies. Yellow and black tiger-striped markings adorn its wings, and it has long "tails" on its hind wings. This beautiful butterfly is often observed gliding lazily along rivers and streams. When disturbed, however, it can flee quite rapidly. Next time you go on a streamside hike, look for the western tiger swallowtail in the wet sand or mud alongside the water. Butterflies sip water from the moist soil to obtain mineral salts and other nutrients that have leached from the surrounding soil and rocks. This is a phenomenon known as "puddling." Male butterflies do more puddling than females. It is thought that the male transfers the dissolved salts and minerals to the female during mating to give her the nutritional boost she needs to lay eggs. Swallowtails are named for the tails on their hindwings that resemble the long tail feathers of swallows.

Western tiger swallowtail

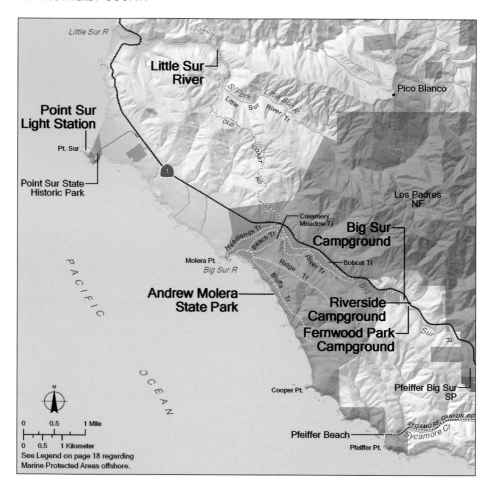

Point Sur

Point Sur
to Big Sur Valley

	Sandy Beach	Rocky Shore	Trail	Visitor Center	Campground	Wildlife Viewing	Fishing or Boating	Facilities for Disabled	Food and Drink	Restrooms	Parking	Fee
Point Sur Light Station			•						•	•	•	
Little Sur River			•									
Andrew Molera State Park	•	•	•	•	•	•	•	•		•	•	•
Big Sur Campground					•					•	•	•
Riverside Campground					•					•	•	•
Fernwood Park Campground					•				•	•	•	•

POINT SUR LIGHT STATION: *Hwy. One at Point Sur, Big Sur Coast.* The light station was built in 1889 on the Point Sur headland, 270 feet above the ocean. The initial light source for its Fresnel lens, since removed, was a whale oil lantern. A modern beacon now operates at the station, which is within Point Sur State Historic Park and is open to visitors on a limited basis. Docent-led walking tours depart from the parking area off Hwy. One on Saturdays at 10 AM and 2 PM and Sundays at 10 AM, weather permitting; additional tours are scheduled in summer, and occasional moonlight tours are offered. No reservations are taken for the tours, which depart promptly and last three hours, requiring a half-mile walk each way and considerable stair climbing. No pets, strollers, or picnicking allowed. Meet at the locked entrance gate on the west side of Hwy. One, a quarter mile north of the former Point Sur Naval Facility. Limited parking; no large RVs. For information, call: 831-625-4419. Visitors to the light station can see surf, rocky coast, and perhaps a whale offshore, but there is no beach access; all the land surrounding Point Sur is private and not open to the public. Point Sur is an island of basaltic rock that is connected to the mainland by a large sandbar; this rare geologic formation is called a tombolo.

LITTLE SUR RIVER: *Off Old Coast Rd., 4 mi. N. of Hwy. One, Big Sur.* Steelhead trout spawn in the undammed Little Sur River. The headwaters of the river support a dense redwood forest. Off Old Coast Rd., a trail leads upstream through the trees into the Los Padres National Forest, beginning one half-mile south of the bridge where Old Coast Rd. crosses the south fork of the river. The private El Sur Ranch borders much of the lower reach of the Little Sur River, including the sandy beach at the river mouth; do not trespass.

ANDREW MOLERA STATE PARK: *Hwy. One, 8.3 mi. S. of Bixby Creek Bridge, Big Sur.* This large state park at the mouth of the Big Sur River offers walk-in camping, hiking on upland trails or along miles of sandy beach, surfing, and horseback riding. The park's vehicle entrance and parking lot are on the Hwy. One, or north, side of the river, while much of the park's land lies south of the river. Summertime plank bridges allow pedestrian crossing when the river's flow is low. The campground and trail to Molera Point are accessible all year, without the need to cross the river.

The campground is located in a meadow, a quarter-mile walk along a trail from the west end of the parking area. There are 24 family campsites, each accommodating up to four persons, with picnic tables, food storage lockers, and firepits. No reservations taken. Restrooms with running water are located near the campground. The campground can also be reached by footpath from a pull-out on Hwy. One, one-quarter mile west of the park's vehicle entrance.

Andrew Molera State Park boasts a spectacularly beautiful sandy beach, one of Big Sur's few surfing areas, located about a mile from the entrance parking area. Beach-

Andrew Molera State Park

goers can cross the Big Sur River near the parking area, then walk along the Beach or Creamery Meadow Trails to a 300-yard-long sandy beach strewn with driftwood and rocky shore farther south. Alternatively, hikers bent on getting to the coast can avoid the river crossing by heading west from the parking area past the campground to Molera Point, covered with wildflowers in spring, where elevated ground provides dramatic views. At Molera Point, however, you are separated from the main beach by the river mouth. Shorebirds such as willets, sanderlings, and red-necked phalaropes frequent the shallow lagoon. Private ranch land extends west from Molera Point; do not trespass.

Other trails in Andrew Molera State Park include the River Trail, which follows the south bank of the Big Sur River; the Ridge Trail, which climbs the spine of Pfeiffer Ridge; and the Bluffs Trail, which parallels the shoreline toward Cooper Point. The Bobcat Trail follows the Big Sur River on its north bank, starting at the entrance parking lot and also accessible from pull-outs on Hwy. One, west of the park entrance. When hiking, watch out for poison oak and check for ticks. Fee for day use and camping; for information, call: 831-667-2315.

The nonprofit Big Sur Historical Society operates a visitor center at Andrew Molera State Park in a historic ranch house, staffed by volunteers and open usually on Saturday and Sunday, 11 AM to 3 PM. The Ventana Wildlife Society conducts bird walks and monarch butterfly walks in the park, participates in the California condor recovery effort, and offers children's summer nature camps; for information, call: 831-455-9514.

The Big Sur Ornithology Lab, a project of the Ventana Wildlife Society, conducts a bird-banding program to monitor the health and viability of migratory songbird populations. Morning bird-banding demonstrations, open to the public, are conducted year-round at Andrew Molera State Park on a variable schedule; call ahead: 831-624-1202. Also at the park, guided horseback rides through western sycamore and redwood trees to the beach are available from Molera Horseback Tours; for information, call: 831-625-5486.

Spanish missionaries at Carmel Mission called the river draining the western slopes of the Santa Lucia Range *El Rio Grande del Sur*, "The Big River of the South." The Big Sur River Valley was homesteaded in the late 1860s by American settlers, including the Pfeiffers and the Posts. The community of Big Sur offers everything, and then some, that a world-famous resort area might be expected to have, from restaurants with ocean views to shops, galleries, and places of spiritual refuge. But a town center there is not; visitor services are strung along Hwy. One, with the largest concentration between Andrew Molera State Park and Pfeiffer Big Sur State Park, and others farther south. The Big Sur Valley is separated from the ocean by Pfeiffer Ridge, ensuring a moderate, sheltered summer climate among the redwoods. Big Sur's beaches are usually cooler than the Big Sur Valley during the summer; beaches are located at several state parks and sites within the Los Padres National Forest.

BIG SUR CAMPGROUND: *Hwy. One, 1.7 mi. N. of Pfeiffer Big Sur State Park, Big Sur.* Tent and RV campsites and cabins near the Big Sur River are available at this privately owned facility. RV sites have hookups. Hot showers, laundry, playground, and store Fee for camping. Open all year; for reservations, call: 831-667-2322.

RIVERSIDE CAMPGROUND: *Hwy. One, 1.6 mi. N. of Pfeiffer Big Sur State Park, Big Sur.* This privately owned campground in the redwoods on the Big Sur River has 28 tent campsites with picnic tables and firepits and 17 sites with electrical and water hookups for RVs up to 40 feet long. Hot showers, laundry, and firewood available. Cabins are also available. Fee for camping; closed during the winter. For reservations, call: 831-667-2414.

FERNWOOD PARK CAMPGROUND: *Hwy. One, .7 mi. N. of Pfeiffer Big Sur State Park, Big Sur.* A privately owned resort on the Big Sur River. Camping facilities include tent cabins, tent campsites, and RV campsites in the redwood forest next to the river; RV sites have water and power hookups. Recreational facilities include a volleyball court and trails connecting to Pfeiffer Big Sur State Park. Also on the property are a motel, restaurant, tavern, and general store. Fee for camping. Open all year; for reservations, call: 831-667-2422.

Andrew Molera State Park

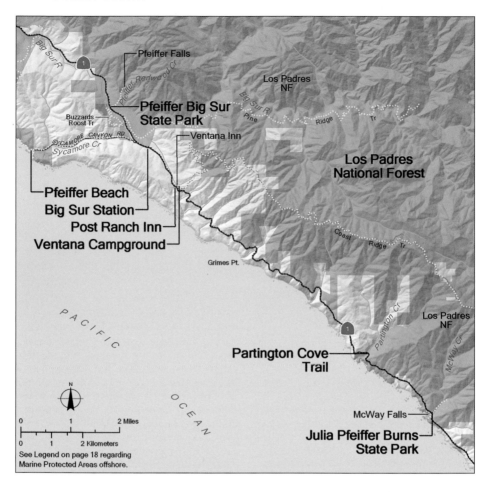

Santa Lucia Range from Pfeiffer Big Sur State Park

Pfeiffer Beach to Julia Pfeiffer Burns State Park

	Sandy Beach	Rocky Shore	Trail	Visitor Center	Campground	Wildlife Viewing	Fishing or Boating	Facilities for Disabled	Food and Drink	Restrooms	Parking	Fee
Pfeiffer Beach	•	•						•		•	•	•
Pfeiffer Big Sur State Park			•	•	•	•		•	•	•	•	•
Big Sur Station			•	•				•		•	•	
Post Ranch Inn				•							•	
Ventana Campground					•				•	•	•	
Los Padres National Forest	•	•	•	•	•	•	•			•	•	•
Partington Cove Trail		•	•									
Julia Pfeiffer Burns State Park		•	•		•			•		•	•	•

PFEIFFER BEACH: *W. of Hwy. One, end of Sycamore Canyon Rd., Big Sur.* It is easy to miss the unsigned highway turn-off to this beautiful, white-sand beach; the intersection is six-tenths of a mile south of Big Sur Station. Narrow Sycamore Canyon Rd. parallels Sycamore Creek. Private property is adjacent; do not trespass. At the end of the road, a sandy trail leads through cypress trees to a beach surrounded by steep cliffs and sea stacks. Waves crash spectacularly through natural arches in the rocks. Hazardous surf. Gusty winds are common; fires are prohibited. A beach wheelchair is available on a first-come, first-served basis when the entrance booth is staffed. Day use only; fee charged. The beach, a unit of Los Padres National Forest, is open daily, 10 AM to 9 PM. Call: 831-385-5434.

PFEIFFER BIG SUR STATE PARK: *Along Hwy. One, 26 mi. S. of Carmel, Big Sur.* This popular state park includes 1,000 acres in the valley of the Big Sur River. The riparian forest includes coast redwood, western sycamore, black cottonwood, and big-leaf maple trees. Trails lead along the valley floor and up the adjacent steep slopes. Pfeiffer Falls on Pfeiffer Redwood Creek is a short walk from the park's nature center. West of Hwy. One, the Buzzards Roost Trail climbs steeply up a forested ridge, offering views of peaks in

Pfeiffer Beach

the Los Padres National Forest. Birds in the park include Steller's jays, canyon wrens, dark-eyed juncos, chestnut-backed chickadees, red-tailed hawks, and belted kingfishers. Coast horned lizards and western tiger swallowtail butterflies also inhabit the park. Nature walks and programs given in the summer.

Pfeiffer Big Sur State Park has 218 family campsites, plus hike or bike campsites and group sites; enroute camping is also allowed. Some sites are wheelchair-accessible. Trailers up to 27 feet long and RVs up to 32 feet long can be accommodated; RV dump station available. For park information, call: 831-667-2315. The Big Sur Lodge, with rooms and cabins, swimming pool and sauna, café, and gift shop, is located within the park; call: 1-800-424-4787.

BIG SUR STATION: *E. of Hwy. One, .6 mi. S. of Pfeiffer Big Sur State Park entrance.* Rangers provide information on Big Sur state parks and the Los Padres National Forest. Fire permits, group picnic area reservations, and wedding permits are also available. A shop offers maps and books. The Pine Ridge Trail starts at the station; leashed dogs allowed on the trail. Open daily, 8 AM to 4:30 PM, year-round; call: 831-667-2315.

POST RANCH INN: *W. of Hwy. One, 2.4 mi. S. of Pfeiffer Big Sur State Park.* Private resort on ridge west of Hwy. One; entrance opposite Ventana Campground. An upland loop trail is open to the public on a permit basis; obtain a pass at the entry kiosk.

VENTANA CAMPGROUND: *E. of Hwy. One, 2.4 mi. S. of Pfeiffer Big Sur State Park.* This privately owned campground shares an entry with the Ventana Inn. Turn east off Hwy. One at the old red house, one-half mile south of the post office, then left to the campground. Eighty campsites with picnic tables and fire rings in the redwoods; running water and hot showers available. Dogs on leash permitted. Campsites are for tents or for RVs up to 22 feet long; no generators allowed. Trails lead up the hill to the Ventana Inn, where there is a restaurant and gift shop. Campground closed in winter; for information, call: 831-667-2712.

Near the Ventana Inn is Cadillac Flat, a starting point for hikers and backpackers using the old Coast Ridge Rd. trail, which winds southeast along the mountains. Park at the vista point sign on the Ventana Inn access road and continue on foot, bearing right to the trailhead. From the Coast Ridge Rd. trail, connecting trails lead down the east side of the coastal range to campsites scattered along the Big Sur River, upstream from Pfeiffer Big Sur State Park.

LOS PADRES NATIONAL FOREST: *Big Sur Coast.* The Los Padres National Forest includes 1.75 million acres of land in the Coast and Transverse Ranges. One part of the National Forest is in Monterey County, including the Big Sur area, and the other spans San Luis Obispo, Santa Barbara, Ventura, and Kern Counties. Ecosystems range from beaches to coastal redwood forests, from mixed evergreen forest to chaparral and oak woodlands. The National Forest includes two California condor sanctuaries, the Sisquoc Condor Sanctuary in Santa Barbara County and the Sespe Condor Sanctuary in Ventura County. Pictographs created by the Chumash indigenous people are found in caves and rock outcroppings. There are numerous picnic locations, campsites, and more than 1,200 miles of trails for day hikers and backpackers.

Wildfires are a concern, and permits are required for campfires or barbecues outside developed campgrounds. Camping outside designated campgrounds is allowed in some areas; contact a ranger station for more information. Pets must be leashed in developed recreation sites; follow posted regulations regarding horses and pack animals. California Department of Fish and Game rules regarding fishing and hunting apply within the National Forest. Off-road vehicles are allowed on designated trails only. Leave natural areas as you found them. Maps, fire permits, and information about Los Padres National Forest are available at Big Sur Station, or call National Forest headquarters in Goleta: 805-968-6640.

PARTINGTON COVE TRAIL: *W. of Hwy. One, 1.8 mi. N. of entrance to Julia Pfeiffer Burns State Park.* At milepost 37.8 on Hwy. One is a trail that leads downhill, across Partington Creek, and through a 200-foot-long tunnel to a former doghole lumber port. (A doghole port, used by ships in 19th-century coastal commerce, is a cove just big enough for "a dog to turn around in.") Forest products, including tanbark and wooden shakes, were transferred to ships here. Partington Cove is now used by divers and sightseers. Park on the shoulder of Hwy. One.

JULIA PFEIFFER BURNS STATE PARK: *E. and W. of Hwy. One, 11 mi. S. of Pfeiffer Big Sur State Park.* Perhaps the most-photographed feature of this state park is the McWay waterfall, which cascades 80 feet onto the beach. Park east of Hwy. One; the Waterfall Trail leads seaward, beneath the road. A marine protected area designation provides special protection to marine resources in state waters adjacent to the state park; for more information, call the Department of Fish and Game: 831-649-2870. Two environmental campsites are located on the headland; turn south from the Waterfall Trail just west of Hwy. One. Campers must register at Big Sur Station; for reservations, call: 1-800-444-7275. The state park also includes hiking trails and picnic areas inland along wood-

The landmarks of Big Sur include the restaurant called Nepenthé, with a dining area 800 feet above the ocean, once known as a hangout for artists and writers during the 1960s. In the redwoods of Castro Canyon is the old-time Deetjen's Big Sur Inn, a rustic resort with redwood cabins built by Norwegian Helmuth Deetjen in the early 1920s. Nearby is the Henry Miller Memorial Library, a community center of sorts as well as a repository of materials on the area's artists and writers, including Henry Miller, who lived on Partington Ridge from 1944 until 1962. Galleries and visitor services are found along Hwy. One.

ed McWay Creek. Up a steeply rising trail is 4,099-foot Anderson Peak, located little more than two miles from the shoreline in the Los Padres National Forest.

California condors, Los Padres National Forest

Julia Pfeiffer
Burns SP

Coast Ridge Tr

Arroyo Seco R.

Ventana
Wilderness

**Esalen
Institute**

Lime Cr

John Little
State Reserve

Los Padres
NF

Coast

Ridge Tr

Big Cr

Landels-Hill
Big Creek Reserve

Gamboa Tr

Cone
Peak

CONE

Gamboa Pt.

Vicente Cr

Cone Peak Tr

San
Antonio
Tr

PEAK

Stone

Ridge Tr

Vicente
Flat

RD.

Lucia

Limekiln

Vicente

Los Padres
NF

Lopez Pt.

Limekiln State Park

Flat Tr

Vicente Flat Trail

Kirk Cr.

NACIMIENTO

Kirk Creek Campground

FERGUSON RD

Mill Creek Picnic Area

Mill Cr

P A C I F I C

O C E A N

N

0   1   2 Miles

0   1   2 Kilometers

See Legend on page 18 regarding
Marine Protected Areas offshore.

Limekiln State Park

Esalen to Mill Creek

	Sandy Beach	Rocky Shore	Trail	Visitor Center	Campground	Wildlife Viewing	Fishing or Boating	Facilities for Disabled	Food and Drink	Restrooms	Parking	Fee
Esalen Institute												•
Limekiln State Park	•	•	•		•		•	•	•	•	•	
Vicente Flat Trail			•		•					•		
Kirk Creek Campground	•	•			•		•		•	•	•	
Mill Creek Picnic Area	•	•	•					•	•	•		

ESALEN INSTITUTE: *W. of Hwy. One, 2.3 mi. S. of Julia Pfeiffer Burns State Park, Big Sur Coast.* The Esalen Institute offers workshops on subjects related to the body, mind, and spirit; personal retreats are also sometimes available. Fees are charged; most services, including massage and use of the natural hot-spring mineral baths, require advance reservations. Call: 831-667-3000.

Between the Esalen Institute and Gamboa Point is the Landels-Hill Big Creek Reserve, which includes 4,200 acres of mountainous terrain and is managed for research by the University of California at Santa Cruz. Very limited opportunities for public access include an annual open house day and participation in occasional extension classes offered by the University or seminars offered by Esalen Institute. For more information, call: 831-667-2543. Hwy. One pull-outs at the Big Creek Bridge and Gamboa Point offer sweeping views of rugged coast.

LIMEKILN STATE PARK: *Off Hwy. One, 2 mi. S. of Lucia.* Limekiln State Park has a sandy beach at the mouth of a steeply sided canyon. Twelve different plant communities, from coastal strand to alpine forest, can be found between the beach and the top of nearby Cone Peak, which rises sharply to an elevation of 5,155 feet. In the park's narrow canyon are 33 campsites, some overlooking the beach and others in a redwood grove by Limekiln Creek. All sites have picnic tables and fire rings. The narrow entrance road limits the creekside campsites to tent camping only; beach sites can accommodate RVs up to 24 feet long and trailers up to 15 feet. Restrooms have showers; firewood is sold at the entrance station. Leashed dogs allowed in campsites only. For camping reservations, call: 1-800-444-7275. From November 1 through mid-March, some campsites are available, on a first-come, first-served basis. For park information, call: 831-667-2403.

The sandy beach is bounded by steep cliffs; pelicans roost on offshore rocks. Surf fishing is popular, but rough water makes swimming inadvisable. Hikers can explore inland trails, each less than a mile long, to a 100-foot-high waterfall or a grove of some of Monterey County's largest redwood trees. The Kiln Trail leads to four historic limekilns constructed in the early 1880s. The kilns, made of stone and steel, were located to take advantage of a limestone deposit. The stone was processed, or "slaked," using heat produced by redwood cut nearby. The resulting lime, used in making cement, was packed and shipped from the mouth of the canyon, then known as Rockland Landing. Stay on trails to avoid poison oak.

VICENTE FLAT TRAIL: *E. side of Hwy. One, 3.9 mi. S. of Lucia.* This mountainous trail starts across Hwy. One from Kirk Creek Campground. Park at pull-outs on the highway; parking in Kirk Creek Campground is for campers only. As you reach an elevation of nearly 2,000 feet on the trail, the broad ocean seems to spread at your feet. The rigorous but scenic climb brings hikers to Vicente Flat, five miles from Hwy. One. Several well-spaced creek-side campsites have picnic tables and barbecue grills. Creek water must be filtered. The Vicente Flat Trail passes through grasslands dotted with spring wildflowers, groves of oak trees, and patches of redwood forest, continuing inland through canyons and to the top of Cone Peak. Primitive camps are located throughout the area. To explore inland trails, hikers can also drive east from Hwy. One for approximately seven miles on winding Nacimiento-Ferguson Rd., then turn north for four miles on unpaved Cone Peak Rd. to a junction with the Vicente Flat Trail. Call: 831-385-5434.

KIRK CREEK CAMPGROUND: *Off Hwy. One, 3.9 mi. S. of Lucia.* There are 33 no-reservation campsites, including hike or bike sites, on a high terrace between Hwy. One and a rocky beach. Campsites have picnic tables and firepits; the campground host sells firewood. Restrooms, not wheelchair accessible, have running water. Fee for camping; trailers and RVs up to 30 feet in length allowed. Call: 831-385-5434.

Dramatic vistas of the Big Sur coast; a rough path leads from the south end of the campground down the hundred-foot-high bluff to the narrow beach, where surf fishing is possible. Campers at Kirk Creek may also use Forest Service–managed Sand Dollar Beach and Pfeiffer Beach with no additional charge. The Nacimiento-Ferguson Rd., which terminates south of the campground, provides a long, winding link to the southern Salinas Valley.

MILL CREEK PICNIC AREA: *W. of Hwy. One, 4.5 mi. S. of Lucia.* A handful of picnic tables with barbecue grills overlook the ocean; a steep path leads to the rocky shore, which is a diving area. Vault toilets. Open all year, 10 AM to 6 PM. Hang gliders who launch from Prewitt Ridge use the blufftop as a landing site. All hang gliders must register with the Pacific Valley Ranger Station; for information, call: 805-927-4211.

Vicente Flat Trail

Hwy. One, Big Sur Coast, 1937

The Big Sur coast has few roads other than Hwy. One, reflecting the area's mountainous character. The Carmel-San Simeon Highway, as Hwy. One through Big Sur was originally called, has been subject to frequent closures, one of them due to the threat of landslides in October 1935, before the road even opened officially.

Hwy. One became part of California's state highway system in 1919, when voters approved issuance of bonds to build the road through Big Sur. Proponents of the new coast road predicted that it would bring visitors, as well as offer better access for the region's residents. Before Hwy. One was built, travelers to the valley of the Big Sur River used the Old Coast Rd., a winding track that follows the north side of Bixby Creek rather than cross its formidable canyon. Early Big Sur resorts that drew visitors over the Old Coast Rd. included Pfeiffer's Ranch, built by the pioneer family who homesteaded in the valley in the 1860s. The ranch resort was eventually acquired in 1934 by the California Department of Parks and Recreation; today, it is Pfeiffer Big Sur State Park.

The completion of Hwy. One dramatically reduced travel time from Monterey to Big Sur and brought more tourist-oriented enterprises to the Big Sur economy of cattle ranching, mining, and lumbering. The 1965 filming in Big Sur of *The Sandpiper*, starring Elizabeth Taylor and Richard Burton, brought more attention. The population of Big Sur has changed little, however, remaining in recent decades at about 1,500. The appearance of the community has changed little, too; under Monterey County's coastal plan, new homes generally must be sited out of view of Hwy. One. In 1965, Hwy. One from the Carmel River south to the San Luis Obispo County line was designated California's first Scenic Highway.

Since the 1930s, the 70-mile stretch of Hwy. One through Big Sur has been closed by natural forces at least 53 times. Four major landslides occurred during the winter of 1983; as a result, Hwy. One was closed for two months north of Big Sur and for more than a year near Julia Pfeiffer Burns State Park. The Big Sur Volunteer Fire Brigade assisted in evacuating those who needed medical attention, while the U.S. Army and National Guard brought in emergency supplies by helicopter. Two weeks after the northern landslide, Hwy. One opened on a limited basis, allowing residents to get in and out by daily convoy. One highway closure takes place on the last Sunday in April for the annual Big Sur International Marathon. Runners start at Big Sur Station and run north on Hwy. One through the Big Sur Valley and along the coast to the Carmel River bridge; call: 831-625-6226.

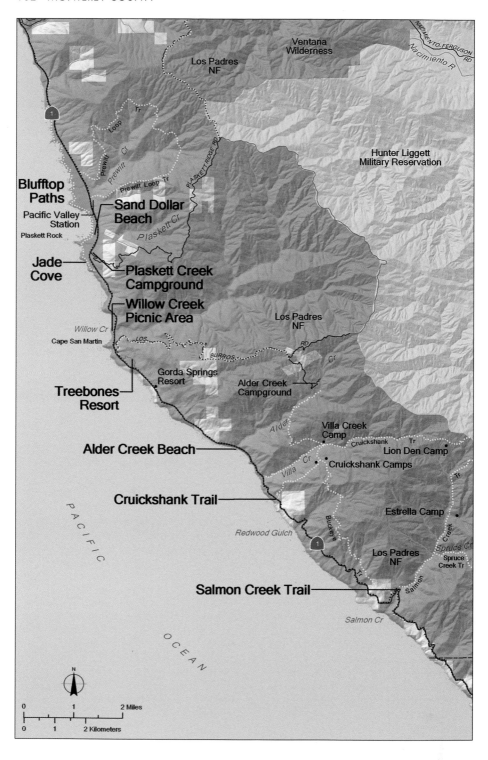

Ventana
Wilderness

Los Padres
NF

Hunter Liggett
Military Reservation

Prewitt Loop Tr

PLASKETT RIDGE RD

NACIMIENTO-FERGUSON RD

Nacimiento R

Prewitt Cr

Loop Tr

**Blufftop
Paths**

Pacific Valley
Station

Plaskett Rock

**Sand Dollar
Beach**

Plaskett Cr

**Jade
Cove**

**Plaskett Creek
Campground**

**Willow Creek
Picnic Area**

Willow Cr

Cape San Martin

LOS

BURROS

RD

Cr

Los Padres
NF

**Treebones
Resort**

Gorda Springs
Resort

Alder Creek
Campground

Alder

Alder Creek Beach

Villa Creek
Camp

Cruickshank   Tr

Lion Den Camp

Villa Cr

Cruickshank Camps

Tr

Cruickshank Trail

PACIFIC

Redwood Gulch

Buckeye

Estrella Camp

Creek

Spruce Cr

Spruce
Creek Tr

Los Padres
NF

Salmon Creek Trail

Salmon

Tr

Salmon Cr

OCEAN

N

0   1   2 Miles

0   1   2 Kilometers

Big Sur South

	Sandy Beach	Rocky Shore	Trail	Visitor Center	Campground	Wildlife Viewing	Fishing or Boating	Facilities for Disabled	Food and Drink	Restrooms	Parking	Fee
Blufftop Paths	•	•	•									
Sand Dollar Beach	•		•			•				•	•	•
Plaskett Creek Campground			•		•					•	•	•
Jade Cove		•	•									
Willow Creek Picnic Area	•	•				•	•			•	•	
Treebones Resort					•				•	•	•	•
Alder Creek Beach			•								•	
Cruickshank Trail			•		•						•	
Salmon Creek Trail			•		•						•	

BLUFFTOP PATHS: *W. of Hwy. One, 6.9 mi. S. of Lucia.* On the broad coastal terrace of Pacific Valley, within Los Padres National Forest, are several paths that lead to the bluff edge for views of the rugged shoreline or, in a few locations, descend to pocket beaches. Look along the shoulder of Hwy. One for stiles that allow hikers to cross the roadside fence. Several stiles are clustered between one and two miles north of Sand Dollar Beach, including one located opposite the U.S. Forest Service's Pacific Valley Station. Another stile is located one-quarter mile south of Sand Dollar Beach. Hikers bent on exploring inland canyons rather than the coast can use the Prewitt Trail; the better-maintained end of the loop trail starts one-half mile north of Pacific Valley Station, on the east side of Hwy. One.

SAND DOLLAR BEACH: *W. of Hwy. One, 11 mi. S. of Lucia.* This crescent-shaped strand is the longest accessible sandy beach on the Big Sur coast. High above the south end of the beach is a picnic area with eight tables and non-wheelchair-accessible restrooms among Monterey cypress trees, and a paved path that leads across a grassy meadow to the edge of the bluff. A steep trail leads down to the shore. To the north, 5,155-foot-high Cone Peak can be seen rising above the Big Sur coast. Pacific sanddab, white croaker, and flounder inhabit the nearshore waters off Sand Dollar Beach. Gulls roost on offshore rocks at the south end of the bay. Hiking, surfing, and surf fishing are popular; this is also a hang glider landing area. The surf can be rough; use caution, as there is no lifeguard on duty. Pets must be leashed. Fee for day use. Part of Los Padres National Forest; call: 831-385-5434.

The coastline north through Pacific Valley to Limekiln State Park and south past Willow Creek is all within Los Padres National Forest. Inland of Hwy. One there are scattered private land holdings among the public lands. National forest lands are open for hiking and exploration although mountain biking and hang gliding are restricted in the Ventana Wilderness area.

PLASKETT CREEK CAMPGROUND: *E. side of Hwy. One, 11.7 mi. S. of Lucia.* The campground has 45 first-come, first-served sites with picnic tables, fire rings, and barbecue grills. Hike or bike sites available. A group campsite for up to 50 persons can be reserved; call: 1-877-444-6777. Drinking water available; restrooms have running water. The campground is set among Monterey pine trees. Open all year; fee for camping. For information, call: 831-385-5434.

JADE COVE: *W. of Hwy. One, .5 mi. S. of Sand Dollar Beach.* Look for the Los Padres National Forest sign on Hwy. One. Shoulder parking only; a steep trail leads down the serpentine cliff face to several rocky coves. A popular diving area; a two-ton jade boulder located underwater has been reported nearby. Call: 831-385-5434.

WILLOW CREEK PICNIC AREA: *W. of Hwy. One, 2.2 mi. S. of Sand Dollar Beach.* This day-use area is located where Willow Creek flows into the ocean. From the vista point at the southeast end of the Willow Creek Bridge, a steep paved road leads down to sea level. The shoreline is rocky to the south of the creek's mouth and sandy to the north. In spite of the rocks, surfers use the beach. Offshore kelp beds provide habitat for harbor seals and southern sea otters. Gulls and cormorants roost around Cape San Martin, a rocky point to the south of the beach. Vault toilets. Open all year, 10 AM to 6 PM.

TREEBONES RESORT: *E. of Hwy. One, off Willow Creek Rd., .9 mi. N. of Gorda.* The right fork of Willow Creek Rd. leads to the privately operated Treebones Resort, which offers five walk-in tent campsites, as well as furnished yurts, a swimming pool, and store; call: 877-424-4787. The left fork of Willow Creek Rd., also known as Los Burros Rd., winds inland eight miles to the Alder Creek Campground in the Los Padres Na-

On the southern Big Sur coast, visitor services are limited. Food and lodging are available at the hamlet of Lucia, near Limekiln State Park, and at Gorda Springs Resort, where gasoline is also available.

tional Forest. Two campsites are set among oak trees along the creek; picnicking and hiking are also possible. Bring your own water. Call: 831-385-5434.

ALDER CREEK BEACH: *W. side of Hwy. One, 2.5 mi. S. of Gorda.* An old road, now interrupted by rockfalls, leads to a rocky beach. A pull-out on Hwy. One offers views of seastacks and the rugged coast.

CRUICKSHANK TRAIL: *E. side of Hwy. One, 3.6 mi. S. of Gorda.* Small roadside pull-outs on both sides of Hwy. One are located at this trailhead; the trail sign, placed parallel to the highway, is easy to miss. The Cruickshank Trail climbs sharply upward for the first mile, quickly reaching 1,500 feet in elevation, with spectacular views over the ocean.

Jade Cove

The trail also includes some gentle sections among trees; overall, the effort required here for a day hiker to explore a couple of miles inland may be somewhat less than on the Salmon Creek Trail, located a few miles farther south.

For backpackers, several campsites are located on the slopes above Villa Creek, within a few miles of Hwy. One. Lower Cruickshank and Upper Cruickshank camps have one site each, with picnic table and firepit. Villa Creek camp, located three miles from Hwy. One, is situated near the creek, with waterfalls and a forested setting. Water must be treated for drinking. The Cruickshank Trail continues inland to Lion Den camp and also connects to other trails, including the Buckeye Trail that leads north to Alder Creek and south to Salmon Creek. South of the Cruickshank Trail is Redwood Gulch, where the coast redwood tree reaches the southernmost limit of its natural range.

SALMON CREEK TRAIL: *E. side of Hwy. One, 8 mi. S. of Gorda.* At Hwy. One milepost 2.2, at a switchback in a steep ravine, is a trail that leads up the slope along the south bank of Salmon Creek. Cascades of water tumble down the stream course, bordered by ferns, alders, and California bay laurel trees. Views of the ocean can be glimpsed through the trees.

After a mile and a half of climbing, hikers on the Salmon Creek Trail reach the Spruce Creek Trail, where the two creeks come together. Spruce Creek camp, at the trail junction, offers a handful of scattered campsites in a forested setting, with swimming holes and waterfalls nearby. A mile and a half farther along the Salmon Creek Trail is Estrella camp, set in a meadow with scattered oak trees. The Spruce Creek Trail leads to Dutra Flat, Turkey Springs, and San Carpoforo camps, the last one eight miles distant from the trailhead at Hwy. One.

The scenic headland known as Plaskett Rock separates Sand Dollar Beach from Jade Cove. If you hike down to Jade Cove, you may be rewarded by finding pebbles of jade in the surf zone; the U.S. Forest Service prohibits collecting above the mean high tide line. The jade, consisting of the minerals jadeite and nephrite, is eroded out of the Franciscan Complex rocks that make up the bluffs in this area. These minerals form at very high pressures in the earth's crust, but at relatively low temperatures. Their presence is proof that parts of the Franciscan Complex rocks were shoved far below the surface during subduction of the Farallon plate.

Salmon Creek

Marble Cone from Pacific Valley

Santa Lucia Range

C ALIFORNIA is known for its complex and active geology. Nowhere is the geologic story more complex than in Big Sur and the Santa Lucia Range. This region, together with Monterey Bay and the San Francisco Peninsula, is separated from the rest of California by two major fault systems. The San Andreas fault borders the region to the east, and the San Gregorio-Hosgri fault system, running mostly offshore, separates it from rocks found to the west (see diagram, page 108).

The Big Sur region is split into two blocks with quite dissimilar geology by yet another major fault, the Sur-Nacimiento fault system. The block lying between the San Andreas and Sur-Nacimiento faults is known as the Salinian block. It contains sedimentary rocks, the Sur Series, that have been intruded by granite masses. The Sur Series has been highly metamorphosed, and it contains schists, gneisses, and marbles. Many of these rocks are resistant to erosion, and they make up the higher peaks of the Santa Lucia Range. Together, these metamorphosed sedimentary rocks and granites are very similar to rocks of the Sierra Nevada, and there seems no doubt that they have been brought up from the south along the San Andreas fault. Lying between the Sur-Nacimiento and the San Gregorio-Hosgri fault systems is the Nacimiento block. This block consists predominantly of Franciscan Complex shales, sandstones, cherts, and basalts. These rocks are found mixed up in what is known as a mélange—a complex jumble of rock types with no clear bedding or stratigraphic order.

To understand how these blocks came to be juxtaposed in the Big Sur area, we need to look back some 130 million years in Earth's history. Today, the North American and Pacific plates are sliding past each other. Big Sur sits firmly on the Pacific plate and is moving northwestward along a series of faults, with most movement along the San Andreas fault. Prior to about 30 million years ago, however, a plate under the Pacific Ocean known as the Farallon plate was colliding with North America rather than sliding past it. The dense rocks of this oceanic plate slid under the less dense continental crust in a process known as subduction. As the Farallon plate was subducted beneath the North American plate, water was introduced into the mantle, lowering the melting temperature of the rocks there and causing them to melt. The resulting magma rose upward into the overlying North American plate, intruding and metamorphosing the sedimentary rocks of the Sur Series. These magma bodies fed a chain of volcanoes similar to the Cascade volcanoes of northern California, Oregon, and Washington. Today, these volcanoes are long eroded away; all that remains are altered remnants of some of the volcanic rocks that they deposited in the Sierra Nevada. Uplift of the Sierra Nevada has exposed the ancient magma masses that fed these volcanoes, now solidified as granite.

At the same time as the volcanic arc of the ancestral Sierra Nevada was active, ongoing subduction of the Farallon plate scraped sediments off the seafloor and piled them on the edge of the North American plate. This accretionary wedge consists of deep sea cherts, shales, some limestones, and basalt, all mixed together in the mélange of the Franciscan Complex. Sediment derived from the North American plate also contributed to the mixture, forming most of the sandstones and shales in the Franciscan Complex.

About 30 million years ago, the last of the Farallon plate in the southern California area was subducted beneath the North American plate. Lying across a mid-ocean ridge was the Pacific plate, which now entered the California area. The plate motion changed from subduction to horizontal sliding, and the plate boundary became what is known as a transform boundary. The Pacific plate moved northwestward relative to North America, sliding along a series of faults, of which the largest is the San Andreas fault. The Sur-Nacimiento and San Gregorio-Hosgri fault systems also played a role, and at times in the last 30 million years actually took up most of the movement between the North American and Pacific plates. These faults sliced obliquely through the volcanic arc and accretionary wedge that were formed previously. Movement along the San Andreas fault brought the Salinian block northward, emplacing it west of Franciscan Complex rocks in central California. Movement along the Sur-Nacimiento fault brought northward the Nacimiento block, consisting mostly of the Franciscan Complex, and emplaced it west of the Salinian block. These two blocks now lie far north of the location where the rocks in them were created.

Exactly how far the Salinian and Nacimiento blocks have migrated northwestward in the last 30 million years is a subject of some debate. There is little doubt that about 300 miles of movement has occurred along the San Andreas fault; well-recognized offsets of geologic features on either side of the fault confirm this amount of movement. Recently, some geologists, relying largely on records of the earth's magnetic field recorded in these rocks, have proposed that the Salinian and Nacimiento blocks migrated from as far away as 1,800 miles, at the latitude of Acapulco, Mexico.

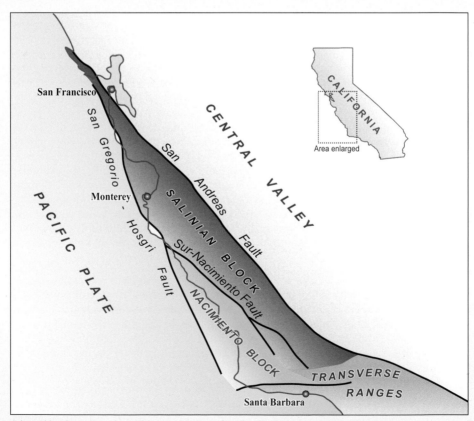

Adapted by Grace Fong from *The Natural History of Big Sur* (Paul Henson and Donald J. Usner)

Santa Lucia Range

Overlying the basement rocks of the Salinian and Nacimiento blocks are sedimentary rocks that record the more recent history of the Big Sur area. Fine-grained shales reflect periods of submergence beneath deep oceans, whereas sandstones were deposited in shallow seas. Scattered coarse conglomerates were deposited in submarine canyons that have now been uplifted. This uplift began about five million years ago, and reflects compression along the boundary between the Pacific and North American plates. Uplift has been episodic, and continues today. Downcutting by streams has not been able to keep up with this uplift, and therefore no gaps exist through the Santa Lucia Mountains, making the interior of this range one of the most rugged and inaccessible in coastal California.

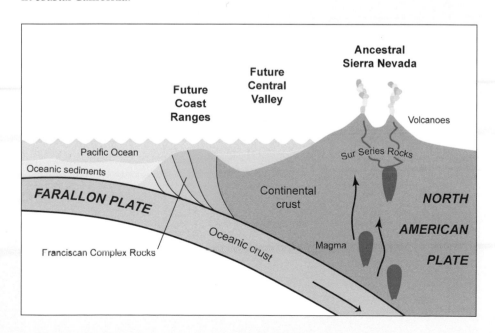

The mountain peaks and deep canyons of the Santa Lucia Range are habitat for the California condor (*Gymnogyps californianus*), one of the largest flying birds in the world. Adult condors weigh about 22 pounds and have a wingspan of up to nine and a half feet. The California condor is also one of the world's most endangered species. Condors were so close to extinction in the mid-1980s that the last 22 wild condors were brought into captivity to establish a controlled breeding program. The condor has been reintroduced to parts of the American West, including the Santa Lucia Range. The updrafts provided by steep gorges provide favorable soaring conditions for the huge birds, and the relative inaccessibility of the Santa Lucia Range makes it a good place to reintroduce this very rare species. Currently there are almost 100 condors in the wild. Yet the condor still hovers on the brink of extinction.

California condors have no feathers on their heads or feet. They are scavengers and feed only on the carcasses of animals. Condors reach sexual maturity when they are six or seven years old, and they are monogamous breeders. Once they reach maturity, condors breed only once every other year. Due to these demographic limitations, the condor population cannot expand very quickly, even under good conditions.

The last wild condors were brought into captivity because of the high risks they faced in the wild, particularly the threat of lead poisoning from scavenging carcasses killed by ammunition. The successful reintroduction of condors to the wild is threatened also by habitat loss, shooting, and collisions with power lines. Nevertheless, if you keep your eye out, you may spy a condor soaring overhead.

California condor

Page opposite: Montaña de Oro State Park, San Luis Obispo County

San Luis Obispo County

MONTEREY

San Antonio
Res

JOLON RD

INTERLAKE
RD

Camp Roberts
Military
Reservation

Nacimiento
Res

NACIMIENTO

LAKE
DR

101

Paso
Robles

46

Inters

Hearst
San Simeon
SHM

Pt. Piedras
Blancas

San Simeon

William Randolph Hearst
Memorial SB

San Simeon SP

Cambria

SAN   LUIS

41

46

OBISPO

Harmony

Atascadero

229

Whale Rock
Res

Pt. Estero

Cayucos

41

58

Inters

Cayucos SB

Los Padres
NF

Santa Margarita

Morro Strand SB

Estero Bay

Morro Bay

Santa Margarita
Res

Morro Bay SP

Morro
Bay

1

S BAY BL

Los Osos

Los Padres
NF

Montaña de Oro SP

LOS OSOS VALLEY RD

San Luis Obispo

PACIFIC

Pt. Buchon

Avila
Beach

227

Lopez L

Pt. San Luis

Pismo Beach

101

O C E A N

Grover Beach

Oceano

Arroyo Grande

Pismo SB

San Luis Obispo

Oceano Dunes
SVRA

1

Bay

Oso Flaco L

To

166

SANTA
BARBARA

135

N

0   10   20 Miles

0   10   20 Kilometers

1

246

San Luis Obispo County

SAN LUIS OBISPO County's northern coast is a pristine landscape of forested mountains carved by perennial streams that wind their way to the sea across grassy marine terraces. Prominent headlands at Ragged Point, Point Sierra Nevada, and Point Piedras Blancas separate long stretches of rocky coast where seabirds nest on offshore rocks, shorebirds forage along the water's edge, and passing whales or playful otters can frequently be seen beyond the surf line. Morro Bay, really an estuary teeming with wildlife that is sheltered from the ocean by a four-mile-long sand spit, lies at the midpoint of the San Luis Obispo County coast. South of Morro Bay the San Luis Mountains, part of the southern Coast Ranges, form a barrier to travel between the bay and the southern county coast. In these mountains, Montaña de Oro State Park provides access to a largely undisturbed landscape with spring wildflower displays, rare geologic formations, and remote canyons where the Chumash indigenous people once walked. The coast highway joins the shoreline again at San Luis Obispo Bay. Continuing south, the coast broadens into a wide plain edged with sandy beaches and undulating dunes from Pismo Beach to the mouth of the Santa Maria River.

The coast of San Luis Obispo County offers a collection of small towns with all the traditional seacoast features, from fishing boats and fishing piers in the harbor to wide, sandy beaches. Cambria features antique shops and miles of dark sand pocket beaches, while Cayucos has the informal charm of the traditional beach community, centered on its wooden pier. The municipality of Morro Bay is of modest size, but the town's efforts to make its waterfront accessible to visitors are dramatically apparent. Many recreational options are contained in a few blocks along Embarcadero Rd., within view of imposing Morro Rock. The Morro Bay Winter Bird Festival is held annually in mid-January, when migratory birds are in abundance around the Morro Bay Estuary; call: 805-772-4467. Surfing or kiteboarding on Estero Bay offer recreational sport to the participants and good viewing for those who choose a beach towel on the sand over the rigors of the waves. Pismo Beach has an old-time, pier-town feel, and the enormously long sweep of sand fronting the community hosts beach activities of every sort.

For visitor information, contact:

San Luis Obispo County Visitors and Conference Bureau, 805-541-8000.

San Simeon Chamber of Commerce, 805-927-3500.

Cambria Chamber of Commerce, 805-927-3624.

Cayucos Chamber of Commerce, 805-995-1200.

Morro Bay Visitors Center & Chamber of Commerce, 805-772-4467.

Los Osos/Baywood Park Visitors Center, 805-528-4884.

Pismo Beach Visitors Information Center, 805-773-4382.

Grover Beach Chamber of Commerce, 805-489-9091.

Oceano Chamber of Commerce, 805-489-2252.

South County Area Transit serves coastal locations in Shell Beach, Pismo Beach, Grover Beach, and Oceano; call: 805-781-4472.

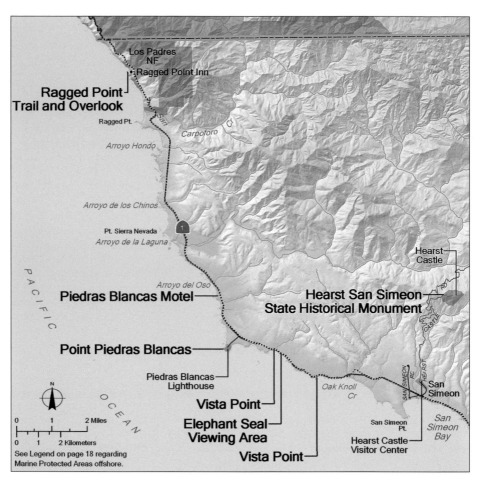

Ragged Point Trail and Overlook

Los Padres NF

Ragged Point Inn

Ragged Pt.

Carpoforo Cr

Arroyo Hondo

San

Arroyo de los Chinos

Pt. Sierra Nevada

Arroyo de la Laguna

1

Piedras Blancas Motel

Arroyo del Oso

Point Piedras Blancas

Piedras Blancas Lighthouse

Vista Point

Elephant Seal Viewing Area

Vista Point

Hearst San Simeon State Historical Monument

Hearst Castle

HEARST RD

SAN SIMEON RD

CASTLE RD

Oak Knoll Cr

San Simeon

San Simeon Pt.

Hearst Castle Visitor Center

San Simeon Bay

PACIFIC

OCEAN

N

0   1   2 Miles

0   1   2 Kilometers

See Legend on page 18 regarding Marine Protected Areas offshore.

Hearst San Simeon State Historical Monument

Ragged Point to Hearst Castle

	Sandy Beach	Rocky Shore	Trail	Visitor Center	Campground	Wildlife Viewing	Fishing or Boating	Facilities for Disabled	Food and Drink	Restrooms	Parking	Fee
Ragged Point Trail and Overlook	•	•	•						•	•	•	
Piedras Blancas Motel	•	•						•	•	•		
Point Piedras Blancas		•										
Elephant Seal Viewing Area	•	•				•				•		
Vista Points		•				•				•		
Hearst San Simeon State Historical Monument				•				•	•	•	•	•

RAGGED POINT TRAIL AND OVERLOOK: *W. of Hwy. One, 15 mi. N. of San Simeon.* Two miles north of Ragged Point, the Ragged Point Inn sits on a level grassy terrace, 300 feet above sea level. Along the bluff edge there are dramatic views of the Big Sur Coast, including the sheer-sided cliffs that plunge down to the surf zone, north of the inn. Behind the restaurant, a very steep, switchbacked trail leads down the north face of the bluff, past a waterfall, to a small sandy beach. The trail and overlook are privately managed; for information, call: 805-927-4502.

PIEDRAS BLANCAS MOTEL: *W. of Hwy. One, 7.3 mi. N. of San Simeon.* The old motel is closed, but beach access is available under interim property management by the Trust for Public Land. The high bluff edge is eroding, but walk 300 yards south of the motel buildings to where the bluff is lower to descend to the sandy beach. Long-range plans

Coast near Ragged Point

From Ragged Point to San Simeon Bay, the Hearst Ranch has dominated the northern San Luis Obispo County coast since the 1860s, when George W. Hearst began to assemble his mammoth land holdings. Under a complex land conservation agreement that became effective in 2005, some 13 miles of Hearst Ranch coastline have changed from private to public ownership. The agreement allowed the State of California to take title to nearly 1,000 acres of land along Hwy. One, including 13 sandy beaches. Additional Hearst Ranch land along Hwy. One will remain under private ownership but will be subject to state control, in order to provide an alignment for completion of the California Coastal Trail through the entire length of the Hearst Ranch. The coastline now under state ownership has become part of San Simeon State Park. Before the newly acquired coastal lands can be opened to public use, however, the California Department of Parks and Recreation must prepare public access and resource management plans. Meanwhile, existing public use will continue of pull-outs and vista points along Hwy. One, maintained by the California Department of Transportation (Caltrans).

As part of the conservation agreement, a permanent easement has been created to allow Caltrans to realign Hwy. One eastward at four locations in the Hearst Ranch, to address continuing coastal erosion problems. The conservation easement also protects existing views of the coast from Hearst Castle and along Hwy. One by restricting future construction. Under the conservation agreement, the land inland of Hwy. One will remain privately owned and not open to the public, except for lands already available for public use, such as Hearst Castle.

The long-standing agricultural and livestock operations will continue inland of Hwy. One, but approximately 80,000 acres of the Hearst Ranch, the largest privately owned working cattle ranch along the coast of California, will come under a conservation easement designed to protect natural and scenic values of the land.

The area subject to the conservation easement adjoins the Los Padres National Forest and includes a wide range of habitat communities, including coastal scrub, chaparral, coast live oak, ponderosa pine forest, riparian, and wetland habitats.

call for the site to be transferred to the State Parks Department; call: 805-927-2020.

POINT PIEDRAS BLANCAS: *W. of Hwy. One, 6 mi. N. of San Simeon.* In 1864 a lookout was built on the point to alert whalers at nearby San Simeon of approaching leviathans during their annual migration. Three centuries earlier, explorer Cabrillo had noted the guano-covered rocks near the point on his voyage up the California coast in 1542 and had called them *piedras blancas,* or "white rocks." The rocks at the point are a California sea lion and harbor seal hauling-out ground, as well as a rookery for Brandt's cormorants.

A lighthouse, originally 115 feet tall, was built on the point starting in 1874 and equipped with a first-order Fresnel lens. (A first-order lens is the largest of seven sizes of lenses in typical use in the 19th century.) When it was built, the Piedras Blancas Lighthouse was the only light between Point Pinos in Monterey County and Point Conception in Santa Barbara County. Originally one of California's few tall lighthouses with a classic silhouette, the Piedras Blancas Lighthouse suffered storm damage in 1949 that resulted in removal of the lantern room at the top of the tower. The original lens and lantern can now be viewed at Pine-dorado Park on Main St. in Cambria. The Piedras Blancas light remains in operation, with a lens installed in 2002. The lighthouse is now owned by the U.S. Bureau of Land Management and is open only for limited public tours; for information, call: 805-927-2968. Tour reservations are handled by the National Geographic Hearst Castle Theater; call: 805-927-6811.

ELEPHANT SEAL VIEWING AREA: *Off Hwy. One, 4.7 mi. N. of San Simeon.* Elephant seals can be seen near Point Piedras Blancas nearly year-round. Caltrans has provided a parking area with interpretive panels at the Elephant Seal Viewing Area, now managed by the California Department of Parks and Recreation. The Friends of the Elephant

Seal, a nonprofit group, organizes a docent program that assists visitors. Volunteers are on site daily, except during September, from 10 AM to 4 PM. The docents, who wear blue jackets, answer questions and provide information about the seals. The group also maintains a visitor center at the Plaza del Cavalier in San Simeon, staffed by volunteers and open variable hours; call: 805-924-1628.

VISTA POINTS: *Along Hwy. One, 5 mi. and 3.5 mi. N. of San Simeon.* Two vista points overlooking beaches and rocky shore are located on the west side of Hwy. One, north of San Simeon. The more northerly of the two vista points has paved parking and is located just north of the Elephant Seal Viewing Area. No facilities.

Elephant seals

Neptune Pool, Hearst San Simeon State Historical Monument

Roman Pool, Hearst San Simeon State Historical Monument

HEARST SAN SIMEON STATE HISTORI-CAL MONUMENT: *E. of Hwy. One on Hearst Castle Rd., San Simeon.* Formerly William R. Hearst's private estate, now open to the public as a State Historical Monument located within the Hearst Ranch holdings. Tours of the monument start at the Hearst Castle Visitor Center, located on Hwy. One. The visitor center has exhibits, a gift shop, food service, and the National Geographic Hearst Castle Theater, where films about Hearst Castle screen continuously; for showtimes, call: 805-927-6811. Visitors to the Castle travel by bus up the hill, and all tours involve a half-mile walk and considerable stair climbing. For tour tickets and advance reservations, which are highly recommended, call: 1-800-444-4445. Fee for tours. For information on accommodations for visitors with mobility limitations, call: 1-800-777-0369.

William Randolph Hearst was a newspaper tycoon known for publishing sensational headlines, opinionated coverage, and accounts that were not necessarily completely accurate. "The Chief" also published popular magazines, pioneered the syndication of comic strips and the production of newsreels, and made movies featuring lavish costumes and sets. He was presumed to be the inspiration for the title character in the film *Citizen Kane*. He involved himself in national politics, and he generated controversy by printing stories that pressured President William McKinley to launch the Spanish-American War in 1898, apparently to increase sales of his newspapers.

William Randolph Hearst's father was mining magnate and U.S. Senator George Hearst, from whom William got his first newspaper, as well as a love for picnicking and camping on the family's extensive land holdings. His mother was Phoebe Apperson Hearst, a philanthropist and lover of culture, through whom he met Julia Morgan, the renowned architect he hired to design and manage the construction of his castle on the 250,000-acre family ranch in San Simeon.

Julia Morgan was one of the first women to earn a civil engineering degree from the University of California at Berkeley, the first woman admitted to the École des Beaux-Arts in Paris, and the first woman to receive a California architect's license (in 1904). She was visionary, hard working, detail-oriented, and willing to accommodate Hearst's collaboration in designing (and often redesigning) his estate on *La Cuesta Encantada*, "the Enchanted Hill."

The Hearst Castle compound includes a palace-like house, the *Casa Grande*, built in the Mediterranean Revival style, accompanied by three guesthouses on an estate of 127 acres of gardens, terraces, pools, and walkways. The main house alone contains 38 bedrooms and 41 bathrooms. Hearst wanted to live in a building that would hold the countless art pieces he collected, and indeed his home is full of valuable paintings and furniture, Greek vases, medieval tapestries, and antique silver.

The grounds are decorated with Greek, Roman, and Egyptian artifacts. His castle integrates art into its very structure as well: new construction accompanies antique ceilings, fireplaces, and mantels he collected from old castles; Persian tiles; ancient mosaics; and windows, columns, doors, and statues from centuries-old buildings in Europe.

Hearst and his companion, movie actress Marion Davies, regularly hosted lively company, including Charlie Chaplin, Cary Grant, Carole Lombard, Charles Lindbergh, President Calvin Coolidge, and many others who were their guests at lavish costume parties and other social occasions.

The guests ate dinner in the grand dining hall that seated more than 60 persons and afterward watched films in the private movie theater. Activities included admiring the animals in the world's largest private zoo, riding horses, playing tennis, and swimming in one of two magnificent pools. Hearst liked to join in the fun when he could take a break from running his 28 newspapers, 18 magazines, 8 radio stations, 2 film companies, and 2 news services.

Northern Elephant Seals

THE MAMMALIAN ORDER *Pinnipedia* (meaning "feather feet") includes eared seals—fur seals and sea lions—and earless, or true seals—harbor seals and elephant seals. Pinnipeds are insulated from the cold by fur, a thick hide, and a fat layer; their limbs have evolved into flippers. Sea lions and fur seals have small external ears, and their hind flippers can turn forward; this is useful on land where they are able to move about quite efficiently. California sea lions are the "seals" seen in circus shows.

Harbor seals and elephant seals cannot turn their hind flippers forward, and are therefore less mobile on land, able only to wriggle along on their bellies. Unlike sea lions and fur seals, true seals float in the water vertically, with their heads sticking out. When ready to submerge, they sink straight down, tail first, rather than diving forward.

Northern elephant seals (*Mirounga angustirostris*) are the largest of the true seals; females are up to 12 feet long, weighing 900 to 1,800 pounds, while males reach 16 feet in length and weigh up to 5,000 pounds. Pups, once weaned, can weigh up to 500 pounds. Males have large, bulbous snouts and rough, dry-looking skin. They can dive to tremendous depths in the ocean, on average between 1,000 and 2,000 feet, occasionally reaching nearly a mile below the surface. Elephant seals feed on fish, squid, and octopus. The seals in turn are prey for white sharks and for orcas, also called killer whales.

Elephant seal social structure is male-dominant; however, rather than defending territories as sea lions and fur seals do, elephant seal males establish a dominance hierarchy among themselves through brief but violent battles. The top-ranking male mates with most females. The breeding season of these animals is from December to March.

Once numerous in the Pacific Ocean, elephant seals were slaughtered in large numbers during the 19th century for their oil-rich blubber. At the low point of their population, only a few dozen elephant seals remained on Guadalupe Island off Baja California.

Elephant seal haul-out near Point Piedras Blancas

Female elephant seal and pup

During the 20th century, the number of elephant seals grew, and the animals established rookeries on the Channel Islands, Año Nuevo Island in San Mateo County, and Southeast Farallon Island off San Francisco. Later, breeding colonies appeared on the mainland. Since 1990, northern elephant seals have established a large breeding colony near Point Piedras Blancas. At first, the colony occupied a small cove near the lighthouse at Point Piedras Blancas. The colony has grown rapidly, both in numbers and in territory, which has expanded to additional beaches along Hwy. One. In 1995, 600 pups were born near Point Piedras Blancas, and the following year, there were 1,000 new pups. In 2005, 3,500 pups were born, and the number of elephant seals at Point Piedras Blancas was estimated at about 14,000, although not all of them are present at any one time. The largest colonies of elephant seals are found at San Nicolas and San Miguel Islands; the total population of the species is estimated at 100,000.

Females give birth from the end of December through February. After pups are weaned, the mothers depart, leaving their offspring to teach themselves how to swim. Adult males also leave the land, spending the next few months feeding at sea. During the spring and summer, adults and juveniles of both sexes return for a month or so to the beaches to molt. After molting, the elephant seals once again take to the ocean to feed. In the fall, the seals haul out on the shore, first the youngest, followed by the mature males at the end of November and females ready to give birth in mid-December. An individual female elephant seal spends a total of two months per year on land, while a male spends four months on land. Elephant seals can be seen along the shore nearly year-round.

When male elephant seals are fighting for dominance, they can move surprisingly fast. Visitors do not want to find themselves in the path of one of these animals that weighs as much as a pickup truck. Elephant seals are protected by law; it is both illegal and dangerous to approach or harass the animals. For more information about the seals, contact the Friends of the Elephant Seal; call: 805-924-1628.

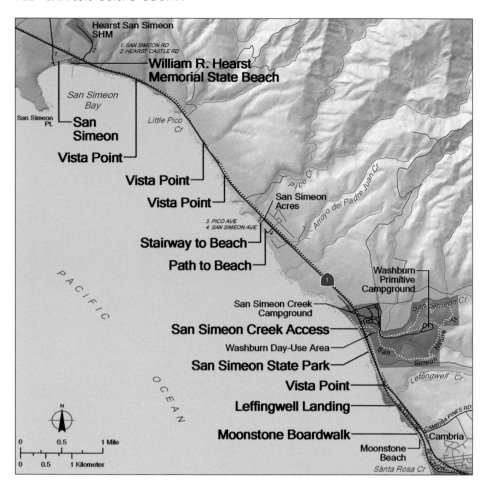

Hearst San Simeon
SHM

1. SAN SIMEON RD
2. HEARST CASTLE RD

William R. Hearst Memorial State Beach

San Simeon Bay

San Simeon Pt.

Little Pico Cr

San Simeon Vista Point

Vista Point

Vista Point

Pico Cr

San Simeon Acres

Arroyo del Padre Juan Cr

3. PICO AVE
4. SAN SIMEON AVE

Stairway to Beach

Path to Beach

PACIFIC OCEAN

Washburn Primitive Campground

San Simeon Cr

San Simeon Creek Campground

San Simeon Creek Access

San Simeon Nature Tr

Washburn Day-Use Area

San Simeon State Park

Leffingwell Cr

Vista Point

Leffingwell Landing

CAMBRIA PINES RD

Moonstone Boardwalk

Cambria

Moonstone Beach

Sànta Rosa Cr

N

0   0.5   1 Mile

0   0.5   1 Kilometer

William R. Hearst Memorial State Beach

San Simeon to Moonstone Beach

	Sandy Beach	Rocky Shore	Trail	Visitor Center	Campground	Wildlife Viewing	Fishing or Boating	Facilities for Disabled	Food and Drink	Restrooms	Parking	Fee
San Simeon									•	•	•	
William R. Hearst Memorial State Beach	•					•	•		•	•		
Vista Points		•									•	
Stairway to Beach	•										•	
Path to Beach	•										•	
San Simeon Creek Access	•		•								•	
San Simeon State Park	•	•	•		•	•		•	•	•	•	•
Vista Point	•										•	
Leffingwell Landing	•	•	•			•	•	•	•	•		
Moonstone Boardwalk	•	•	•					•				

SAN SIMEON: *Hwy. One, 6.5 mi. N. of Cambria.* The village was a whaling station in 1864 before it was acquired by mining magnate George W. Hearst, later a U.S. Senator representing California. Hearst acquired several Mexican land grants reaching from San Carpóforo Creek south to San Simeon Creek, eventually amassing a quarter-million acres of land. Hearst purchased the land after the drought of 1863–64 devastated livestock and forced many of the rancheros to sell land to pay their debts. At San Simeon, Hearst built warehouses and a pier, from which he exported tallow, hides, grain, and quicksilver from cinnabar mines on the ranch land. Later, exports included local dairy products, dried seaweed, and abalones. The Sebastian General Store in San Simeon, little altered since it was established in the 1870s, offers the visitor a glimpse of life here more than a century ago.

WILLIAM R. HEARST MEMORIAL STATE BEACH: *W. of Hwy. One on San Simeon Rd., San Simeon.* A long curve of sandy beach is located east of wooded San Simeon Point. The point provides shelter from waves, making the water good for swimming. The day-use park includes picnic facilities in a eucalyptus grove and in the grassy area near the park entrance. Picnic tables, barbecue grills, and restrooms with running water. A 1,000-foot pier allows ocean fishing; no license necessary for pier fishing, although catch limits apply. Barred surf perch are caught all year, with greater numbers in December and January. Rainbow, rubberlip, and walleye perch are also taken. Deep-sea fishing from chartered boats is also available. Kayaks and bodyboards are for rent near the beach. Call: 805-927-2020.

VISTA POINTS: *Along Hwy. One, .7, 1.8, and 2.4 mi. S. of San Simeon.* Three pull-outs are located on the west side of Hwy. One. Great views of the rocky coast; no facilities.

STAIRWAY TO BEACH: *W. end of Pico Ave., San Simeon Acres.* At the end of the Pico Ave. cul-de-sac, just south of Pico Creek, is an overlook with benches and a short stairway leading to a long sand and cobble beach. Park on the street; no facilities.

PATH TO BEACH: *End of San Simeon Ave., San Simeon Acres.* A wide, packed-gravel path to the beach is located at the mouth of Arroyo del Padre Juan Creek, seaward of Hearst Dr., the Hwy. One frontage road. The path is located on the south side of the Cavalier Inn. On-street parking; no facilities.

SAN SIMEON CREEK ACCESS: *W. of Hwy. One at San Simeon Creek, 2 mi. N. of Cambria.* A long sandy beach, part of San Simeon State Park, extends north and south of the

Leffingwell Landing, San Simeon State Park

Hwy. One bridge over San Simeon Creek. A long pull-out on a low bluff has picnic tables with great views of the coast. The beach is a popular surfing spot. Restrooms and additional parking are at the nearby Washburn Day-Use Area, located on the inland side of Hwy. One and linked to the beach by a path under the highway. The shoreline extending more than two miles south of San Simeon Creek is part of the state park; the sandy beach is interrupted by rocky promontories. The ocean off recently expanded San Simeon State Park is part of the Monterey Bay National Marine Sanctuary, where boating, fishing, surfing, and diving are encouraged, but personal motorized watercraft are not allowed. Other restrictions apply; for information, contact the entrance station at San Simeon State Park or the Monterey Bay National Marine Sanctuary office at 831-647-4201. A Marine Sanctuary visitor center opened in San Simeon in 2006.

SAN SIMEON STATE PARK: *2 mi. N. of Cambria, between San Simeon Creek and Santa Rosa Creek.* San Simeon State Park includes shoreline, coastal scrub habitat, streams bordered by riparian woodlands, and one of the world's few native stands of Monterey pine trees. The San Simeon Nature Trail leads inland from the Washburn Day-Use Area past seasonal wetlands to higher ground, where there are views of the ocean and the Santa Lucia Range. Part of the trail is wheelchair accessible. Along the three-mile-long loop trail look for western bluebirds, western meadowlarks, and raptors such as red-shouldered and red-tailed hawks. Other wildlife in the park includes bobcats, coyotes, brush rabbits, and coastal black-tailed deer; monarch butterflies overwinter in the park. San Simeon State Park includes archaeological sites of human habitation that are nearly 6,000 years old. Junior ranger programs and summertime campfire talks are offered. For park information, call: 805-927-2020.

There are two campgrounds at San Simeon State Park. The San Simeon Creek campground, located close to the beach and dotted with trees, offers 115 sites with picnic tables and fire rings for tent camping or RVs up to 35 feet long. Restrooms with running water, coin-operated showers, and RV dump station available. Firewood can be purchased from the campground host. The Washburn Primitive campground, located on a plateau one mile inland, has 68 sites with fire rings and picnic tables; running water available. Restrooms and some campsites are wheelchair accessible. Campgrounds are open year-round; reservations can be made for dates between March 15 and September 30; call: 1-800-444-7275. Dogs must be leashed during the day and kept in a tent or vehicle at night; dogs are not permitted on trails or on the beach. In addition to the campgrounds and San Simeon Creek beach access point, San Simeon State Park includes day-use areas at Leffingwell Landing, Moonstone Beach, and Santa Rosa Creek, and the shoreline that was formerly part of Hearst Ranch.

VISTA POINT: *N. end of Moonstone Beach Dr., off Hwy. One, Cambria.* An overlook offers a view of the long sandy beach, part of San Simeon State Park. A path leads north 300 yards to a beach access stairway.

LEFFINGWELL LANDING: *Moonstone Beach Dr., .2 mi. S. of intersection with Hwy. One, Cambria.* This day-use unit of San Simeon State Park has picnic tables among the cypress trees, a coarse sand pocket beach, and tidepools. Restrooms have running water. This is a good spot for perhaps spying a sea otter resting off the rocky shoreline. The California Sea Otter Game Refuge includes the ocean and shoreline here; do not disturb the otters. Leffingwell Landing, with its sheltered beach, is a favorite place to launch a kayak or small boat by hand. Rockhounding on state beaches is allowed, although no more than 15 pounds per person per day may be taken.

MOONSTONE BOARDWALK: *From Leffingwell Landing to Santa Rosa Beach, Cambria.* A wheelchair-accessible boardwalk runs one mile along the low ocean bluff, beginning at the south side of Leffingwell Landing and overlooking rocky shore and dark-sand pocket beaches. Dogs on leash allowed on the boardwalk. Stairs along the way provide access to the beach. Park along the shoulder of Moonstone Beach Dr.

Moonstone Beach, that part of San Simeon State Beach extending from Leffingwell Creek to Santa Rosa Creek, is well known among rock hounds for the variety of beautiful pebbles that can be found there. The pebbles are mostly eroded from the Franciscan Complex rocks making up much of the headwaters of Santa Rosa, Leffingwell, and San Simeon Creeks that carry the pebbles to the shore. Here they are rounded and polished by the surf, and cast up onto the beaches during storms. Most of the pebbles are varieties of chert—a lustrous mineral consisting of microscopic grains of silica and derived from the skeletons of tiny marine plankton known as radiolarian.

Colors of the Coast

CALIFORNIA'S Central Coast is sometimes blanketed with coastal overcast, rendering a gray appearance to the landscape, but visitors who look closely will see a rainbow of colors in the plants, animals, rocks, and even the sea itself. Sometimes subtle, sometimes striking, each color of the coast has a story to tell.

Indian pink

RED

Indian pink (*Silene californica*) is a member of the carnation family. It has fire-engine-red blossoms with fringed petals. The stunning red flowers appear between May and June. Indian pink sprouts a circle of floppy stems from a central root system. The plant appears in a variety of habitats, including fully exposed rocky places in grasslands and in oak woodlands. If you spy an Indian pink in full regalia, pause for moment; the flowers have a reputation for attracting hummingbirds, and if you are lucky you may hit the jackpot of colors.

Sand verbena

Sand verbena (*Abronia umbellata*) is a succulent perennial with brilliantly colored pink flowers. The genus name *Abronia* comes from the Greek word *abros*, meaning "graceful or delicate." Sand verbena is a member of the coastal strand plant community and is typically found on beaches and sand dunes. It is quite a sight to see, this vibrant splash of color among the tans and grays of the dunes. Sand verbena is well adapted to the salty environment; its fleshy leaves and stems allow it to store water. It also has a deep taproot that serves as a reservoir for food and water supplies.

Garibaldi

ORANGE

Garibaldi (*Hypsypops rubicundus*) is a beautiful orange marine fish that makes its home in rocky areas and kelp forests near the shore. The name "garibaldi" comes from the Italian patriot, Giuseppe Garibaldi; his forces wore bright red shirts. The garibaldi is one of the few fishes known to use the same nesting site every year. In the spring, the male garibaldi returns to his previous nesting site and spends about a month rebuilding and tending his nest. The female then comes in search of the best nest site, and if the male is a good housekeeper, the female may choose his nest.

Orange-colored **ancient stream deposits**, here eroded into towering pinnacles by coastal erosion, owe their color to iron in the sediments. These sediments consist mostly of sand and gravel laid down in near horizontal layers by streams flowing toward the coast between one and several hundred thousand years ago. Later uplifted and exposed to erosion, minerals in the sediments broke down chemically, a process continuing today. The breakdown of iron-bearing minerals releases the iron and exposes it to oxygen, which together form a series of new minerals yielding the orange hue, a process much like rusting of iron. Layers containing more iron-bearing minerals take on stronger, deeper colors.

Ancient stream deposits

YELLOW

The **coast fiddleneck** (*Amsinckia spectabilis*) is a native annual plant with bright yellow flowers and hairy stems that mostly lie flat on the ground and curve up at the tips. The graceful curved flowers are quite a contrast to the prickly stems. Found on coastal dunes, the coast fiddleneck blooms from April to August. The name "fiddleneck" refers to the plant's flowering stem, which resembles the neck of a fiddle. California's Miwok Indians ate the young leaves of the fiddleneck and ground the seeds to make meal.

Coast fiddleneck

Wilson's warbler (*Wilsonia pusilla*) is a small and sprightly, brilliant yellow neotropical migrant. It spends the winter in the south, from Mexico to Panama, and breeds in streamside thickets from California to Alaska. The male Wilson's warbler has a distinct black crown, while the female has an olive crown. Wilson's warblers can be found in areas near water and are often present in willows and thickets. The birds have a distinctive and cheerful song of rapid, chatter-like notes: *chi chi chi chi chet chet*. Consider taking a walk along a riparian woodland this spring to listen for this colorful and chatty little warbler.

Wilson's warbler

Pacific tree frog

GREEN

The **Pacific tree frog** (*Hyla regilla*) is a common and cosmopolitan frog. Members of this species are small in size and may be any color from tan to bronze or bright emerald green. The Pacific tree frog has a conspicuous dark mask extending from the nostrils through the eyes. This frog is one of our loudest amphibians. During breeding season, males will call to attract females. The advertisement call is a loud, two-part *kreck-eck* with the last syllable rising. This call can be heard from as far away as a mile, and it attracts females from near and far.

Bay pipefish

The **bay pipefish** (*Syngnathus leptorhynchus*) is related to the Pacific seahorse. The common name "pipefish" was applied because of its similarity to the long, slim pipes men smoked in the 1700s. Bay pipefish are usually found in association with eelgrass in intertidal areas. Pipefish have small tubular mouths, and they feed on live crustaceans. To eat, a hungry pipefish gets its tiny, toothless mouth up close to its prey and then takes a big gulp. Like seahorses, the female pipefish places her eggs in a pouch on the underside of the male's body, and the male incubates the eggs. The bay pipefish is truly a unique intertidal fish.

Western bluebird

BLUE

The **western bluebird** (*Sialia mexicana*) is a delicate little bird with a charming personality. The bluebird's song is a short and subdued *cheer, cheer-lee, churr*. With its brilliant bright-blue color and sweet song, the bluebird is a cheerful resident of the woodlands. The western bluebird is a secondary cavity nester; however, bluebirds will also use a birdhouse, and many people have success in attracting them into their yards with nest boxes. The western bluebird prefers open oak woodlands as well as coniferous forests and riparian woodlands. If you have suitable habitat on your property you might consider putting up a bluebird house.

Why is the **ocean** blue? The common response, that the sea reflects the blue of the sky, is only part of the explanation. Even when the sky is not being reflected, the ocean appears blue because water absorbs rays at the red end of the visible spectrum of light, leaving red's complementary color, blue. Shallow water looks pale blue, while deep water absorbs more light and takes on a darker, indigo hue. The ocean's color can be modified by suspended particles such as algae, which create a green color. Whitecaps tend to lose their blue color entirely because they consist largely of tiny bubbles that keep light from penetrating enough to produce a blue tint.

Ocean

VIOLET

Blue dicks (*Dichelostemma capitatum*), a delightful member of the lily family, can be found blooming in many coastal plant communities between February and May. Blue dicks have purple bracts, which hold the tight heads of purple funnel-shaped flowers in terminal clusters. The plants are most commonly observed in grasslands, where they provide a burst of violet in a sea of green grass. The leaves are long, wispy, and grass-like, accenting the elegance of this native plant. Blue dicks grow from an onion-like bulb, and therefore they are long-lived and will bloom again and again in the same spot.

Blue dicks

The **California blackberry** (*Rubus ursinus*) flowers in March and April and produces small flavorful purple blackberries by late May and June. The dark purple berries are delicious, but avoid the red ones, which are too sour. This native vine grows in moist soils, generally near a water source. The California blackberry provides a favorite source of food for birds, coyotes, and people. The species name *ursinus* means "bear" and is a testimony to the ability of these berries to attract bears.

California blackberry

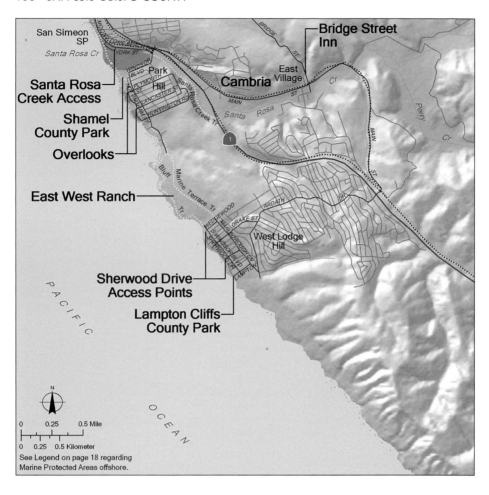

Santa Rosa Creek, San Simeon State Park

Cambria Area

	Sandy Beach	Rocky Shore	Trail	Visitor Center	Campground	Wildlife Viewing	Fishing or Boating	Facilities for Disabled	Food and Drink	Restrooms	Parking	Fee
Santa Rosa Creek Access	•		•			•	•			•	•	
Shamel County Park	•						•			•	•	
Overlooks		•										
Bridge Street Inn								•	•	•	•	
East West Ranch	•	•	•				•				•	
Cambria	•	•	•	•		•	•	•	•	•	•	
Sherwood Drive Access Points	•	•									•	
Lampton Cliffs County Park		•									•	

SANTA ROSA CREEK ACCESS: *W. of Hwy. One, near S. end of Moonstone Beach Dr., Cambria.* At the mouth of Santa Rosa Creek, which marks the southern end of the California Sea Otter Game Refuge, is a parking area with benches overlooking a wide, sandy beach strewn with driftwood. The lagoon near the mouth of the creek is habitat for resident herons and migratory grebes, ducks, and shorebirds. The beach is popular with surfers and bodyboarders. A dirt path leads north along the blufftop, connecting to the wheelchair-accessible Moonstone Boardwalk. Part of San Simeon State Park; call: 805-927-2020.

SHAMEL COUNTY PARK: *Windsor Blvd. at Nottingham Dr., Cambria.* Both a neighborhood park and an ocean access point, this site has a grassy playing field sheltered by trees along with group and family picnic sites, a children's playground, and a swimming pool. Two stairways and a ramp lead to a wide sandy beach, south of the mouth of Santa Rosa Creek. Parking is along Windsor Blvd. and at the south end of the park.

OVERLOOKS: *W. of Nottingham Dr. at Plymouth and Lancaster Streets, Cambria.* Views of rocky shoreline are available on two undeveloped parcels of land on Nottingham Dr.,

View of Cambria from East West Ranch

On Saturday, September 9, 1769, a band of Spanish soldiers, Indians, and an engineer named Miguel Costansó camped at a spot they called *El Estero*, or "the estuary," apparently where Villa and Ellysly Creeks disembogued. The nearby point of land and bay still retain the name of Estero. Led by Juan Gaspar de Portolá, Governor of Baja California, the group had set out overland on July 14 from San Diego, site of California's first mission and presidio. Their goal: to find the Bay of Monterey, described in extremely favorable terms by explorer Sebastián Vizcaíno, 167 years earlier.

Costansó's diary describes a route that generally followed the coast. Mid-September found the party near today's Cambria, where at Pico Creek Costansó wrote that the explorers enjoyed abundant watering places, fine pasture for the animals, and firewood. After a few more days, however, the expedition reached the sheer cliffs north of Ragged Point, forcing a strenuous inland route over the Santa Lucia Range to the Nacimiento and Salinas River Valleys, eventually bringing the group to Monterey Bay, which they did not recognize.

After proceeding as far as San Francisco Bay, the Portolá Expedition returned along much the same route, living on game and gradually running low on food supplies. They were sustained at many places, however, by hospitable Indians, who provided them with acorns, seeds, and fish. On December 24 they were again near Cambria, at Santa Rosa Creek, where they were visited, and fed, by some 200 natives, who were given glass beads in return. On January 24, 1770 the Portolá Expedition returned to San Diego, having endured scurvy, hunger, and hardship on their round-trip journey of nearly 1,000 miles. On a subsequent trip the following spring, Portolá and his company established a mission and presidio at Monterey. Mission San Carlos Borroméo was moved later to the mouth of the Carmel River.

opposite the ends of Plymouth St. and Lancaster St. Dirt paths lead to the bluff edge; there are no facilities.

BRIDGE STREET INN: *4314 Bridge St., Cambria.* This hostel in a century-old house is located in the East Village of Cambria, a mile from Moonstone Beach. Private rooms and bunkrooms are available; shared bathrooms. A kitchen and living room are available for the use of guests. No smoking; no pets. For reservations and information, call: 805-927-7653, between 5 and 9 PM.

EAST WEST RANCH: *Along Hwy. One, between the Park Hill and West Lodge Hill neighborhoods of Cambria.* East West Ranch is an open space and recreation area fronting on the ocean within the community of Cambria. The property's shoreline can be reached from both the north and south sides via Windsor Blvd., which dead-ends on either side of the property. The north stub of Windsor Blvd. ends south of Huntington Rd. in the Park Hill neighborhood. The south stub of Windsor Blvd. ends north of Wedgewood St. in the West Lodge Hill neighborhood. Connecting the two ends of Windsor Blvd. on the East West Ranch property is a mile-long blufftop boardwalk and packed-earth path overlooking the rugged shoreline. Benches are located along the path; gray whales and dolphins may be spotted offshore, along with sea otters in the kelp beds. The trail is wheelchair accessible. Unimproved paths lead down the bluff to tidepools.

The separate Marine Terrace Trail, for hikers and bicyclists, runs inland from the bluff, parallel to the coast, starting at Marlborough Ln. near Wedgewood St. Other trails lead inland, up the ridge to a stand of native

Monterey pine trees. The Santa Rosa Creek Trail runs along the riparian corridor on the west side of the creek; park at the Cambria Community Services District wastewater treatment plant west of Hwy. One, on Windsor Blvd. Santa Rosa Creek provides habitat for the tidewater goby, steelhead trout, and red-legged frog.

Dogs must be leashed on the Bluff Trail; on other trails on East West Ranch, dogs are allowed if under voice control at all times. Small parking areas are located at both trailheads on Windsor Blvd.; no facilities. Other trails crisscross the property; a trail link to the beach from near Main St. in Cambria is planned. Funds for permanent protection of the 430-acre property, originally operated as the Fiscalini Town Ranch, came from a variety of sources, including the State Coastal Conservancy, American Land Conservancy, Cambria Community Services District, and community groups. The open space area is managed by the Cambria Community Services District, assisted by the nonprofit entity known as North Coast Small Wilderness Area Preservation. Call: 805-927-6223.

CAMBRIA: *On Hwy. One, 6.5 mi. S. of San Simeon.* William Leffingwell built a saw mill on Leffingwell Creek about 1861, which would supply Cambria with lumber for the next two decades. A year later cinnabar ore, from which quicksilver, or mercury, is extracted, was discovered in the Santa Lucia Mountains east of the town, and a mining boom resulted. Copper, quicksilver, dairy products, and cattle hides were loaded onto schooners at nearby San Simeon Bay until Leffingwell built a pier in the cove at Cambria in 1874. In 1894 a rail line from San Luis Obispo was extended to Cambria, ending the coastal shipping trade. The town remained an isolated outpost until 1937 when the coast highway was finally completed between Cambria and Carmel. Cambria offers lodging, eating establishments, shops, and numerous galleries in a setting dominated by native Monterey pine forest.

SHERWOOD DRIVE ACCESS POINTS: *W. of Sherwood Dr. at the ends of Wedgewood and Harvey Streets, Cambria.* Access down a low bluff to the sandy beach and rocky shore is along Sherwood Dr. at the ends of Wedgewood and Harvey Streets. At Wedgewood St. there is a bench overlooking the ocean and a wooden stairway down to the beach. Limited parking at both locations, which are in a residential neighborhood. The beaches are composed of coarse sand and pebbles; tidepools are abundant among the rocks.

LAMPTON CLIFFS COUNTY PARK: *Sherwood Dr. at Lampton St., Cambria.* At the south end of Sherwood Dr. is a small county park overlooking the shoreline. Parking is at the end of Windsor Blvd. There are benches, paths, and a stairway down to a rocky reef. Call: 805-781-5930.

Sherwood Drive Access Point

Coastal Prairie

GRASSLAND is one of those major plant community types that we all can recognize instantly without any scientific training. Prairies and tree-studded savannas are found throughout the world, covering perhaps 20 percent of the earth's land surface. The native grasslands of California fall into two broad categories: northern coastal prairie and valley grassland. Northern coastal prairie is scattered along the coast from Monterey County north to Oregon; within this area, grassy fields and bluffs were originally populated by bunchgrasses such as oatgrass and the graceful purple needlegrass. Valley grassland was most common in the Central Valley and includes purple needlegrass, as well as other perennial species such as the sod-producing creeping wildrye.

Although the coastal prairie is defined by perennial bunchgrasses, many associated annual species provide a striking spring wildflower display. Now often a subtle beauty, these wildflowers must once have painted whole sections of the coast with reds, purples, and yellows. The keen eye may still find patches of such unrestrained floral exuberance. Hidden within the grassy thatch are populations of rodents that wax and wane with the weather. Coastal grasslands are home to larger mammals as well, and to numerous species of birds.

Coastal prairie

California plantain (*Plantago erecta*), or dotseed plantain, is a diminutive annual plant, less than six inches tall, with long, soft hairs. The leaves cluster at the base of the plant, forming a basal rosette. California plantain flowers between March and May. The papery flowers are very small and have an almost silvery color. The genus name *Plantago* comes from the Latin *planta*, meaning "footprint." When Europeans came to North America, they introduced another species from the genus *Plantago*, and the Native Americans called it "white man's footsteps." California plantain hosts a variety of butterfly larvae, including the endangered bay checkerspot. The caterpillars eat the leaves of the small plant until they are ready to pupate.

California plantain

Creeping wildrye (*Leymus triticoides*) is a rhizomatous, or mat-forming, native grass with blue-green leaves that grow from two to four feet tall. The graceful flowers grow in upright spikes up to eight inches long. Flowering begins in late spring, and mature seed is present by mid-summer. Due to its spreading nature, a stand of creeping wildrye may span a large area, excluding the growth of other grassland plant species. After the annual grasses have set seed and turned brown, creeping wildrye with its beautiful foliage will not be difficult to recognize. It may be one of the only grasses in the grassland maintaining color into the summer, likely due to the fact that it likes to grow in low-lying and wet areas.

Creeping wildrye

Before producing its blooms, the **soap plant** (*Chlorogalum pomeridianum*) can be recognized by its attractive wavy leaves. This delicate member of the lily family (*Liliaceae*) produces basal leaves arising from a scaly bulb and produces a tall, leafless flower stalk in mid-summer. However, the fragrant white flowers of soap plant are seldom observed in full bloom because they are vespertine, meaning that they open at dusk. They are pollinated by night-flying insects such as small moths. This plant was of considerable value to Native Americans who used the natural saponins (soaps) of this onion relative as a shampoo, a topical treatment for dermatitis, and as a fishing agent (the crushed leaves stupefy the fish).

Soap plant

Yellow mariposa lily

The **yellow mariposa lily** (*Calochortus luteus*) starts life as a single slender leaf followed by flowering stems bearing several blooms in succession from late spring to early summer. Yellow mariposa lilies have three yellow, fan-shaped petals that form a cup up to three inches across. The base of the petals is variously marked with dark brown lines or large splotches. *Mariposa* is the Spanish word for butterfly, and the markings inside the cup of the flower do indeed look like a butterfly. This plant has the ability to reproduce asexually by means of small "bulblets" in the leaf axils, which drop to the ground and grow into new plants. Mariposa lily corms, similar to bulbs, were gathered by Native Americans and baked for a tasty treat.

Owl's clover

In the spring whole fields may be awash in the pinkish-purple color of **owl's clover** (*Castilleja densiflora*). It is a very common annual plant of grasslands, open fields, and serpentine soils. The colorful parts of the owl's clover are actually the small, leaf-like pink bracts; the tiny flower is located within the bracts. The individual flower vaguely resembles an owl, with lobes forming a yellow beak and "eye-spots" on the lower lip. Owl's clover is a "root-parasite," meaning that it gets some of its nourishment by growing into the roots of other plants.

Turkey vulture

Often seen soaring overhead, the **turkey vulture** (*Cathartes aura*) rides on upcurrents of warm air, called thermals, as it scans the ground for a potential meal. Unlike most birds, turkey vultures have a keen sense of smell. They use this ability to help them locate carrion (dead animals), their primary source of food. As they soar, they hold their wings in a slight V and rock drunkenly from side to side. The scientific name for turkey vulture means "cleansing breeze." It got its common name because its bald red head was thought to resemble that of a male turkey. The bald head is a great adaptation for a bird that eats dead animals. Turkey vultures save energy by lowering their body temperature at night and then basking with wings outstretched to warm back up in the morning. They play an important role in nature by cleaning up dead animals from the forests and fields.

Western meadowlarks (*Sturnella neglecta*) are denizens of the open country. They look distinctively different from other members of their taxonomic group, the blackbird family. The meadowlark is easily recognized by its long bill, bright yellow chest, and distinct black collar. Have you heard the song of the western meadowlark? It is a series of flute-like gurgling or bubbling notes that sound like *shee-oo-e-lee shee-ee le-ee*. This memorable song, produced by the males, is performed to attract females and is a proclamation to other males that the territory is occupied. The distinctive flute-like yodel can be heard throughout the grasslands. Look for this brightly colored singer perched on a fence post or tall plant as you enjoy a walk in coastal grasslands.

Western meadowlark

The **broad-footed mole** (*Scapanus latimanus*) is a subterranean tunnel dweller. It spends about 99 percent of its time underground. Moles are not rodents; they belong to the group of mammals known as *Insectivora*. They eat insects and other invertebrates just below the ground surface. Moles have large powerful front paws, designed perfectly for digging, tiny little ears embedded in their fur that allow no dirt to enter, and eyes the size of pinpricks. In addition, their fur is napless, allowing the broad-footed moles to move either forward or backward in their tunnels. These are all interesting adaptations that have evolved to allow the mole to live its underground lifestyle comfortably.

Broad-footed mole

The **coyote** (*Canis latrans*) is a member of the dog family. One difference between dogs and coyotes is that the coyote's round and bushy tail is carried low when running, below the level of its back, while dogs carry their tails high. The genus name *Canis* means "dog" in Latin, and the species name *latrans* means "barking." The common name, coyote, is from the Aztec word *coyotl*, which also means "barking dog." Coyotes are considered one of the most vocal of North American mammals. Their mournful howls and yapping barks fill the night with haunting songs. Howling and barking are used to communicate the position or hunting success of an individual or to reinforce the social bonds of a pack. If you are out on a moonlight hike, consider calling back to see if the coyotes answer.

Coyote

Cayucos Beach

Estero Bluffs to Morro Strand

	Sandy Beach	Rocky Shore	Trail	Visitor Center	Campground	Wildlife Viewing	Fishing or Boating	Facilities for Disabled	Food and Drink	Restrooms	Parking	Fee
Estero Bluffs	•	•	•									
Cayucos State Beach	•						•	•		•	•	
Cayucos Beach	•									•	•	
Morro Strand State Beach (North)	•							•		•	•	
North Point Natural Area	•	•								•		
Morro Strand State Beach Campground	•				•			•		•	•	•
Morro Strand State Beach (South)	•	•						•		•	•	

ESTERO BLUFFS: *From Villa Creek to N. Ocean Ave., Cayucos.* This relatively new state park acquisition includes four miles of blufftop trails through rolling grasslands, with access to beaches and creeks. Seasonal closures at the north end of the park near Villa Creek are necessary to protect endangered snowy plovers. Please obey signs. Dogs must be kept on leash while on the trail and are not allowed on beaches.

The oceanfront Sea West property, west of the hamlet of Harmony, is an even newer state park acquisition. Opening of the park to public use awaits provision of parking and other facilities. For information, call: 805-927-2020.

CAYUCOS STATE BEACH: *Foot of Cayucos Dr., Cayucos.* A wide, sandy beach is located in the pleasant small town of Cayucos; the name means "kayak" in Spanish. This is a popular place for sunbathing, swimming, surfing, bodyboarding, and beach activities. The wooden fishing pier is lit at night and is wheelchair accessible. Group barbecue and picnic facilities are in the patio area adjacent to the Veterans Memorial building, at the foot of the pier. There is parking along Ocean Front Ave. and in the lot north of the pier. Lifeguards are on duty during summer months; a beach wheelchair is available. The San Luis Obispo County Parks Department maintains the beach; call: 805-781-5200.

A Junior Lifeguard program is offered during the summer for strong swimmers, age 10 to 17. Working with U.S. Lifesaving Association instructors, participants learn about rescue techniques and the ocean environment. For information on the Junior Lifeguard program, call: 805-781-5930.

On N. Ocean Ave. at Cayucos Dr. is the redwood Victorian residence built between 1867 and 1875 by Captain James Cass, who also built the original 940-foot pier that was a regular stop for the 19th century Pacific Coast Steamship Company's ships. The classic beach town of Cayucos still has an old-time feel. The community celebrates the Fourth of July with a sand sculpture contest, parade, barbecue, and evening fireworks show. Call: 805-995-1200. Since 1968, the Abalone Farm mariculture operation near Cayucos has raised abalone for restaurants and home cooks; call: 805-995-2495.

CAYUCOS BEACH: *Along Pacific Ave. between 1st and 22nd Streets, Cayucos.* Nine stairways along Pacific Ave. lead to the sandy beach below; all are marked "public." Private property adjoins both sides; do not trespass.

MORRO STRAND STATE BEACH (NORTH): *Along Studio Dr., S. of 24th St., Cayucos.* The sweep of sandy beach facing Estero Bay continues from Cayucos to Morro Rock. Morro Strand State Beach includes three linear miles of beach. Take the 24th St. exit off Hwy. One and bear left to reach a paved parking lot, picnic tables, and restrooms located next to the sand. An additional unpaved parking area is nearby, at the north end of Studio Dr.

Morro Strand State Beach

Farther south on Studio Dr., which is a residential street, there are nine signed paths to the beach located between houses. For information about Morro Strand State Beach, call: 805-772-2560.

NORTH POINT NATURAL AREA: *N. end of Toro Ln., W. of Hwy. One, Morro Bay.* Outside the entrance to the Morro Strand State Beach Campground, turn right (north) on Toro Ln. to reach a city of Morro Bay day-use parking area. There are steps leading down to a beach dominated by rocky outcroppings. No facilities. Call: 805-772-6278.

MORRO STRAND STATE BEACH CAMP-GROUND: *End of Yerba Buena St., W. of Hwy. One, Morro Bay.* Three miles of Morro Strand State Beach extend south from the campground. The long stretch of sand is popular for strolling, fishing, windsurfing, and kite flying. The park offers a 104-site campground for RVs up to 24 feet in length. Campsites can be reserved from Memorial Day to Labor Day and are available on a first-come, first-served basis the remainder of the year. No electric or water hookups available. Fee for camping. For reservations, call: 1-800-444-7275.

MORRO STRAND STATE BEACH (SOUTH): *Ends of Azure St. and Atascadero Rd., W. of Hwy. One, Morro Bay.* Day-use parking and restrooms are available at the west end of Azure St. Take the San Jacinto St. exit off Hwy. One and jog left on Coral Ave. A sandy path leads seaward through the dunes, and a paved path leads south inland of the dunes. The city of Morro Bay's Cloisters Park, with a paved pathway leading through wetlands and dunes, is located west of Coral Ave., south of Azure St. Another Morro Strand State Beach day-use parking area is located at the south end of Embarcadero Rd., off Atascadero Rd.; no restrooms at this location. The sandy beach is backed by dunes and dominated visually by the mass of Morro Rock. For park information, call: 805-772-2560. The breakers north of Morro Rock are a popular surfing location. An annual invitational contest, known as the Big, Bad, and Ugly Surf and Turf, is held here during the winter by the Estero Bay Surf Club; longboard surfing is followed by golf competition. At 1700 Embarcadero Rd., across from Morro Strand State Beach, is the Morro Dunes Campground, a privately operated facility for RVs, trailers, and tents. For reservations, call: 805-772-2722.

Morro Bay

The Morro Bay Power Plant with its 450-foot-high stacks competes in prominence with nearby Morro Rock as a landmark on the San Luis Obispo County coast. The Power Plant has a large influence on not only scenic views, but also the economy in the small community of Morro Bay. The plant's original power generation units were installed by Pacific Gas and Electric in the 1950s; the plant was sold to Duke Energy in 1998 and then to LS Power Group in 2005. Boilers in the plant produce steam, which drives turbines that in turn power the electrical generators. To cool the steam after it passes through the turbines, up to 725 million gallons per day of seawater are drawn from the Morro Bay estuary near the northern T-Pier. Tunnels into the ocean, near Morro Rock, provide an outlet for the cooling water, which is allowed by regulatory agencies to be no more than 20 degrees Fahrenheit warmer than the ambient seawater temperature. The four generating units of the plant have a combined electrical production capability of 1,002 megawatts. Designed originally to run on both fuel oil and natural gas, the plant has used only relatively clean-burning natural gas since 1995. In recent years, due to its relatively old and less-efficient components, the plant has operated only intermittently.

In 2004, the California Energy Commission approved a thorough upgrade of the Power Plant that promises to reduce its impact on the environment while providing more electrical energy than before. New, more efficient generators will produce 1,200 megawatts of energy, while reducing the amount of seawater used for cooling by about half. Many of the existing oil tanks will be removed. As for the tall stacks that are so prominent on the skyline, these will be replaced with four new stacks measuring 145 feet in height.

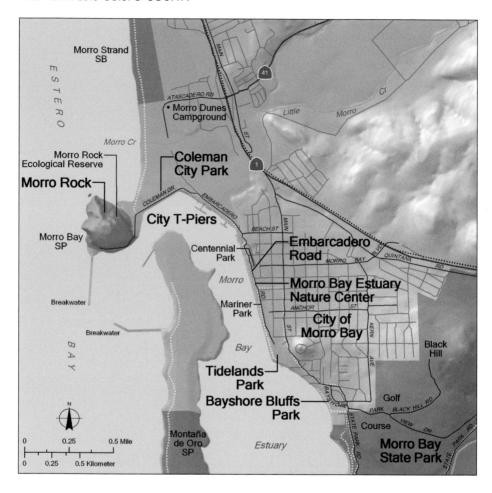

Kayak rentals, Morro Bay

Morro Bay

	Sandy Beach	Rocky Shore	Trail	Visitor Center	Campground	Wildlife Viewing	Fishing or Boating	Facilities for Disabled	Food and Drink	Restrooms	Parking	Fee
Coleman City Park										•	•	
Morro Rock	•						•			•	•	
City T-Piers							•	•	•	•	•	
Embarcadero Road				•			•	•	•	•	•	
Morro Bay Estuary Nature Center				•				•				
City of Morro Bay	•	•	•	•	•	•	•	•	•	•	•	
Tidelands Park							•	•		•	•	
Bayshore Bluffs Park										•	•	
Morro Bay State Park	•		•	•	•	•	•	•	•	•	•	•

COLEMAN CITY PARK: *N. of Coleman Dr., Morro Bay.* This park near the base of Morro Rock features playground equipment and a skateboard park, as well as picnic tables and non-wheelchair-accessible restrooms. The beach northwest of Coleman City Park is a popular surf spot.

MORRO ROCK: *W. end of Coleman Dr., Morro Bay.* In 1542 explorer Juan Rodríguez Cabrillo named the distinctive dome-shaped rock *El Moro.* While humans use the promontory as a navigational aid, peregrine falcons depend on the steeply sided landmark for nesting sites. In 1970 California's population of peregrine falcons dropped to only ten, as a result of eggshell thinning and consequent fledgling mortality caused by pesticide contamination. After DDT was banned in 1972 and measures were taken, such as the captive breeding program of the Santa Cruz Predatory Bird Research Group, the population of these predators has revived. The peregrine falcon remains a protected species in California, however, and the Morro Rock Ecological Reserve is off-limits to climbing. Fishing, wildlife viewing, and sightseeing at the base are permitted. The breakwater on the southwest side can be hazardous during heavy swells. Parking and restrooms are near the end of Coleman Dr. For information on the Ecological Reserve, call the Department of Fish and Game: 707-944-5500.

CITY T-PIERS: *N.W. end of Embarcadero Rd., Morro Bay.* The two piers are used for fishing and commercial boat docking. Restaurants and parking are on or near both piers; the North Pier also has bait and tackle, restrooms, and a public shower. U.S. Coast Guard boats berthed there are sometimes open by appointment for tours; call: 805-772-1293. For boating information, call the harbormaster: 805-772-6254.

EMBARCADERO ROAD: *Along the waterfront, W. of downtown Morro Bay.* The street ends along the Embarcadero between Beach and Anchor Streets provide public access for bay viewing and fishing. Among the restaurants and commercial visitor facilities are pocket parks, including Centennial Park, which has a giant chessboard, picnic tables, and restrooms; and Mariner Park, which offers picnic tables and bay access. Call: 805-772-6278. Kayaks, electric boats, and bicycles can be rented on Embarcadero Rd., where surf shops and bait and tackle suppliers are located and deep sea fishing trips can be arranged.

MORRO BAY ESTUARY NATURE CENTER: *601 Embarcadero Rd., Suite 11, Morro Bay.* Displays about efforts to protect the estuary and watershed are open to the public next to the Morro Bay National Estuary Program office. The Nature Center has a stunning view of Morro Rock. Open daily, 10 AM to 5 PM; call: 805-772-3834.

Morro Rock...the most striking scenic feature of the coast of California.

—Pioneering California geologist
H. W. Fairbanks, 1904

Whether or not you agree with Fairbanks's superlative, there is no question that the enormous mass of Morro Rock, nearly 600 feet high, dominates the landscape of Morro Bay. Upon close examination, the rock can be seen to be made up of interlocking crystals characteristic of igneous rocks—those rocks derived from molten magma. The rock consists of rather large crystals set in a matrix of much finer material. The large crystals indicate slow cooling, which gave the crystals time to grow, but the finer matrix indicates much faster cooling. Igneous rocks formed by such two-stage cooling processes are characteristic of volcanoes, in which magma is brought up quickly from depth to near the surface. Morro Rock is the remnant of a magma body that lay beneath and fed an ancient volcano some 20 million years ago. The volcano itself, and all of the rocks of the Franciscan Complex into which the magma intruded, has long since been eroded away, leaving the plug-like dome of Morro Rock.

A line of similar igneous plugs extends to the southeast for some 18 miles. The somewhat erroneously named "Nine Sisters" (Black Hill, Cerro Cabrillo, Park Ridge, Hollister Peak, Cerro Romualdo, Chumash Peak, Bishop Peak, Cerro San Luis Obispo, and Islay Hill) actually represent about 13 large intrusions similar to Morro Rock. At one time, they fed a line of volcanoes that probably grew along a fracture in the earth's crust. Volcanic rocks and remnants of ancient volcanoes occur sporadically along the California coast, reflecting the subduction of the Farallon Plate beneath the North American continent. As the triple junction between the North American, Pacific, and Gorda Plates migrated up the coast to its present position off Cape Mendocino, leakage of magma through the remains of the spreading center left a trail of volcanoes, decreasing in age to the north. The Morros are examples of these.

Morro Rock was once an island, with ocean inlets on the south and north sides of the rock. Before the turn of the 20th century, Morro Rock began to serve as a quarry for hard basaltic material that was used in breakwater construction. By 1911, a land bridge built out to the rock became complete, replacing the north entrance to the estuary and forming a "tombolo"—the term for a spit that connects an island to the mainland.

CITY OF MORRO BAY: *12.5 mi. N.W. of San Luis Obispo.* Two wharves existed for the export of local dairy and ranching products when the town of Morro Bay was established in 1870. Lumber schooners landed in the bay, bringing redwood and pine from Santa Cruz County. By the turn of the 20th century Morro Bay, like many central and southern California coastal villages, was attracting vacationers who came by train or car to camp on the sandy beaches. In the 1940s Morro Bay began to develop as a commercial fishing port, and it became one of the more important fishing centers along the California coast. Commercial fishing continues in Morro Bay, although at reduced levels that reflect declining fish stocks and short seasons; the catch includes Dungeness crab, prawns, albacore, and salmon.

TIDELANDS PARK: *S. end of Embarcadero Rd., Morro Bay.* This city park has a two-lane boat ramp with docks, a picnic area with barbecue grills and restrooms, and a nautical-themed playground. Fish-cleaning station near the ramps; parking for boat trailers. For information, call: 805-772-6278.

Morro Rock and igneous plugs to the southeast

BAYSHORE BLUFFS PARK: *Bayshore Dr., Morro Bay.* This city park offers access to the estuary's edge, along with picnic facilities, barbecue grills, and restrooms.

MORRO BAY STATE PARK: *S. end of Main St., Morro Bay.* Morro Bay State Park, which includes frontage along the Morro Bay estuary as well as uplands to the east, offers an unusually broad array of visitor attractions. The park campground, renovated in 2005, is located on State Park Rd. in a forest setting. There are 135 family campsites, some wheelchair accessible, for tents or RVs; two group camping areas; and enroute campsites. For reservations, which are required for the group sites, call: 1-800-444-7275. Hot pay showers are available, and there is an RV dump station.

Day-use facilities include picnic tables, barbecue pits, and a trail that overlooks the large marsh along Chorro Creek. For park information, call: 805-772-7434. Morro Bay State Park also includes an 18-hole golf course with pro shop and café on the hills overlooking Morro Bay and the ocean; follow Black Hill Rd. to the summit.

Bays and Estuaries

B AYS AND ESTUARIES form where fresh and salt water mingle at the mouths of rivers. These tidally influenced areas support salt marsh, mudflats, and sometimes eelgrass beds. They are biologically diverse and provide important rearing habitat for many ocean creatures. Plants and animals that occur in estuaries must be able to cope with fluctuations in water depth and salinity; indeed, many of the organisms that occur there have remarkable adaptations. If you take a trip to Morro Bay or an estuary at the mouth of one of the Central Coast's rivers, take the time to look closely at the insects, birds, and plants that are readily observed. You may notice one of the many remarkable ways that members of this biological community have evolved to deal with these challenging conditions.

Sweet Springs Nature Preserve, Morro Bay

Alkali heath (*Frankenia salina*) is a low, sprawling, bushy shrub that is commonly found in the drier parts of the salt marsh. Alkali heath is a member of the sea heath family (*Frankeniaceae*), a taxonomic group of halophytes, or plants that are adapted to growing in salty areas. Salt marshes are generally rich in nutrients but are subjected to fluctuations in acidity, oxygen concentrations, and salinity. Salt marsh plants succeed in this environment where most others cannot compete because they have adapted to these extreme and variable conditions. Alkali heath has small, succulent leaves that have the ability to store water and excrete salt. Look for the plant's small, solitary, rose-purple flowers that bloom between June and October.

Alkali heath

Saltmarsh bird's beak (*Cordylanthus maritimus*) is an annual plant with alternate bluish-green, hairy leaves. Saltmarsh bird's beak has purplish white flowers with a distinct shape that reminds one of a little bird's beak pointing upward. Saltmarsh bird's beak is semi-parasitic, meaning that it attaches its roots to those of other plants, such as pickleweed and saltgrass, in order to obtain water and minerals. The plant is listed by both state and federal agencies as endangered due to habitat loss. Look for saltmarsh bird's beak in coastal salt marshes, just above the high tide line. It blooms from May to October.

Saltmarsh bird's beak

The **tidewater goby** (*Eucyclogobius newberryi*) is a small fish, rarely exceeding two inches in length. Its diet is made up of small crustaceans, aquatic insects, and mollusks. Reproduction occurs year-round, although there are spawning peaks in April and May. When breeding, the males dig burrows in which the females deposit their eggs. Tidewater goby are found only in California and are restricted to brackish waters of coastal lagoons, where they spend their entire lives. The tidewater goby occurred historically in over 100 coastal lagoons in California, from San Diego to Humboldt County. It has since disappeared from most of these sites. The decline of the tidewater goby can be attributed to upstream water diversions, dredging, pollution, siltation, and urban development. The tidewater goby was listed as endangered by the U.S. Fish and Wildlife Service in 1994.

Tidewater goby

Bat ray

The **bat ray** (*Myliobatis californica*) swims gracefully by flapping its bat-like "wings," actually pectoral fins. Bat rays are found living close to the shores of bays and sloughs. They prefer to live in areas with sandy or muddy bottoms where they have easy access to food. When feeding, the bat ray swims along the bottom searching for evidence of clams or other prey. Then it vigorously flaps its wings to expose its sand- or mud-dwelling meal. Once the prey is exposed, the bat ray uses its snout to dig it up. This process ends up creating a large pit, which allows smaller fish access to their prey species. Some small fish rely on this relationship with bat rays because they are not able to dig their own food out of the sand. The bat ray's whip-like tail is as long as its body and has from one to three venomous barbed spines at the base. However, these docile animals use their barbed spines only to defend themselves.

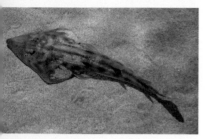

Shovelnose guitarfish

Its pointed snout and guitar-shaped body give the **shovelnose guitarfish** (*Rhinobatos productus*) its name. The shovelnose guitarfish is in the same family as sharks and rays, and it is adapted to life on the sand or mud. Its coloration is cryptic, meaning well camouflaged, and the fish can blend into the sandy habitat. Buried in the sand or mud with only its eyes sticking out, the shovelnose guitarfish lies in wait for an unwary crab or flatfish to pass by. Suddenly, the sand erupts, the guitarfish ambushes its prey, and a meal is consumed.

Spiny dogfish shark

The **spiny dogfish shark** (*Squalus acanthias*) occurs in shallow, sandy habitat and has sharp, venomous spines in front of each dorsal fin. The dogfish shark uses the spines defensively by curling around in a bow to strike an enemy. The genus name *Squalus* is Latin for "a kind of sea-fish" while the species name *acanthias* translates as "a prickly thing," describing the spines found adjacent to the dorsal fins. Spiny dogfish are quite aggressive and have a reputation of relentlessly pursuing their prey. They are called dogfish because they travel and hunt in packs. Gathered together, they sweep an area, eating the fishes in front of them, including herring, sardines, anchovies, and smelts.

Measuring about the size of a sparrow, the diminutive **California black rail** (*Laterallus jamaicensis cotorniculus*) is the smallest North American rail. The black rail is secretive and is more often heard than seen. It typically spends its time skulking in dense vegetation along the edge of coastal salt and brackish marshes. If you are lucky, you may hear its distinctive call, a repeated *kic-kee-doo* or *kic-kic-kerr*, most frequently voiced at dawn or dusk during the early breeding season, from late April to mid-May. Black rails nest in or along the edge of a marsh, usually in tall grasses. The nest is usually concealed in a clump of vegetation, with plants arched over it so that it is hidden from above. The black rail is listed as a threatened species in California due to the loss and fragmentation of wetland habitat.

California black rail

The **American bittern** (*Botaurus lentiginosus*) is a secretive, medium-sized heron with a stout body, short legs, and a white neck. The cryptic plumage and elusive behavior of American bitterns enable them to dwell, often undetected, within densely vegetated marshes. When disturbed, the bittern assumes a reed-like position in which the head points skyward and the body sways back and forth, as if it were a blade of grass blowing gently in the breeze. Listen for the call, *pump-er-wink*, during dusk or dawn hours in the springtime. The eerie call, for which the species is well known, has won it many nicknames, including water-belcher, stake-driver, thunder-pumper, and mire-drum. The species is relatively unstudied due to the birds' secretive nature and inaccessible habitats.

American bittern

Among dabbling ducks, one of the most distinctive in appearance is the **northern shoveler** (*Anas clypeata*). Its bill is large and spatulate, or shovel-shaped. Northern shovelers prefer shallow saltwater or freshwater wetlands. The shoveler feeds by swimming through the water with its bill submerged, drawing water into the bill, and then pumping it out through the long comblike lamellae that line the edges of the bill. In this way the bird can filter from the water small food items, such as aquatic animals, plants, and seeds. Northern shovelers are drawn to areas where they see other birds feeding. They take advantage of the food particles stirred up by the other birds swimming or wading in the area.

Northern shoveler

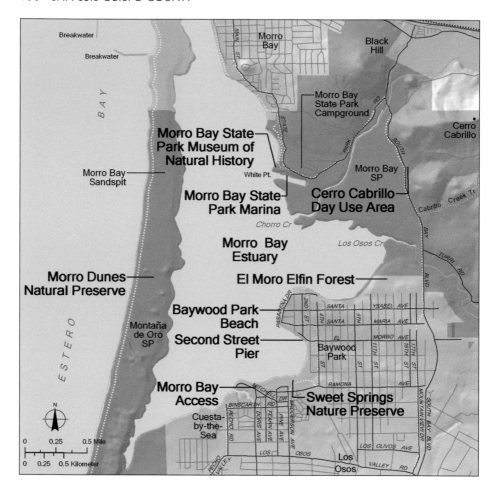

Morro Bay State Park Museum of Natural History

Morro Bay and Baywood Park

	Sandy Beach	Rocky Shore	Trail	Visitor Center	Campground	Wildlife Viewing	Fishing or Boating	Facilities for Disabled	Food and Drink	Restrooms	Parking	Fee
Morro Bay State Park Museum of Natural History				•		•		•	•	•	•	•
Morro Bay State Park Marina							•		•	•	•	
Morro Bay Estuary	•					•	•					
Cerro Cabrillo Day Use Area			•								•	
El Moro Elfin Forest			•			•		•			•	
Baywood Park Beach	•						•				•	
Second Street Pier							•				•	
Sweet Springs Nature Preserve			•			•		•			•	
Morro Bay Access						•	•					
Morro Dunes Natural Preserve	•		•			•						

MORRO BAY STATE PARK MUSEUM OF NAT-URAL HISTORY: *State Park Rd. at White Point, Morro Bay.* The museum is sited on a promontory overlooking the estuary and Morro Rock beyond. Interactive exhibits describe the natural and cultural history of the Morro Bay area. A native garden contains plants used by the Chumash people. Free entry for those 16 years of age and under; open 10 AM to 5 PM daily, except Thanksgiving, Christmas, and New Year's Day. The Museum, which is wheelchair accessible, is operated by the Central Coast Natural History Association and the California Department of Parks and Recreation; for information, call: 805-772-2694. North of the museum are eucalyptus and cypress trees, where from February to June, double-crested cormorants, great egrets, and great blue herons nest, easily visible and audible from the edge of the grove.

MORRO BAY STATE PARK MARINA: *State Park Rd., E. of White Point, Morro Bay.* Morro Bay State Park includes a small marina, where canoes and kayaks can be rented or launched. There is also mooring space for sailboats and power craft, as well as a café.

MORRO BAY ESTUARY: *Hwy. One, 13 mi. N.W. of San Luis Obispo.* The estuary at Morro Bay is the Central Coast's most extensive complex of wetlands. The habitats of the estuary attract many bird species, and in win-

ter the bay is teeming with birds. Particularly common are migratory waterfowl such as pintails, green-winged teals, lesser scaups, wigeons, ruddy ducks, and buffleheads. Shorebirds are abundant along the bay margins, and species such as marbled godwits, willets, and sandpipers number in the thousands. Other shorebirds seen here are curlews, dunlins, dowitchers, and sanderlings. Terns, pelicans, grebes, and loons are also winter visitors. Wading birds—herons and egrets—inhabit the bay year-round.

Great egret, Morro Bay Estuary

El Moro Elfin Forest

Numerous marine invertebrates live in the mudflats of Morro Bay. Mud shrimp, lugworms, and fat innkeeper worms share their burrows with crabs and segmented worms. Washington clams and geoducks are collected in the mudflats; shells found in nearby Native American middens indicate that at one time gaper and bent nosed clams, basket cockles, littleneck clams, and native oysters also were found here.

The Morro Bay National Estuary Program is a federally sponsored effort that seeks to maintain the estuary's water quality, its populations of shellfish, fish, and wildlife, and its recreational activities. Goals of the program include slowing the process of sedimentation, which causes the estuary to gradually become shallower; re-establishing habitat in Chorro and Los Osos Creeks for steelhead trout; and maintaining existing commercial shellfish-raising operations in the estuary. The program maintains a visitor center on the waterfront in the town of Morro Bay; call: 805-772-3834.

CERRO CABRILLO DAY USE AREA: *Off South Bay Blvd., 1.2 mi. N. of Santa Ysabel Ave., Morro Bay.* Part of Morro Bay State Park is located east of South Bay Blvd., overlooking the marsh where Chorro and Los Osos Creeks converge. Trails starting at the Cerro Cabrillo Day Use Area lead to rock-climbing areas and the tops of Cerro Cabrillo and Park Ridge, two of the igneous mounds known

as the Nine Sisters. A second trailhead with a very small parking area is located three-quarters of a mile north of Santa Ysabel Ave.; look carefully for the abrupt turnoff.

EL MORO ELFIN FOREST: *South Bay Blvd. and Santa Ysabel Ave., Baywood Park.* This 90-acre nature preserve can be entered from the northern ends of 11th through 17th Streets; the main entrance and wheelchair access to the boardwalk is at 16th St. A boardwalk winds in a loop around the property. Plant specimens are numbered; pick up a trail guide at the entrance to learn more about them.

The property includes a forest of diminutive oak trees on slopes overlooking the Morro Bay Estuary. Look for lace lichen festooning the branches of the 200-year-old coast live oak trees. Stunted in their growth by the coastal dune environment, some are only 12 feet tall.

The coastal dune scrub, which acts to stabilize the sand, includes California sagebrush and black sage; look for white-crowned sparrows perched on the shrub tops. Maritime chaparral plants such as chamise can also be seen here. The Small Wilderness Area Preservation organization leads guided nature walks on the third Saturday of the month at 9:30 AM, starting at the north end of 15th St.; for recorded information, call: 805-528-0392. At the north end of 3rd St., an overlook maintained by the Morro Coast Audubon Society provides a view of the estuary.

BAYWOOD PARK BEACH: *W. side of Pasadena Dr., Baywood Park.* A public path leads from the small parking area on Pasadena Dr. to a small bay beach with benches overlooking the water. The beach provides access to Morro Bay mudflats at low tides. There is also a bay-viewing platform, located near the end of Pasadena Dr., north of Baywood Park Beach.

SECOND STREET PIER: *S. end of Second St., Baywood Park.* A short T-pier over shallow water provides views of the Morro Bay estuary. Small boats can be launched from the beach here.

SWEET SPRINGS NATURE PRESERVE: *Ramona Ave. W. of 4th St., Baywood Park.* A quarter-mile-long boardwalk leads through a grove of trees to a spot looking out over the estuary. Abundant bird life includes shorebirds, ducks, brant, scaup, wigeon, and snowy egret. The trees harbor overwintering monarch butterflies. The preserve is managed by the Morro Coast Audubon Society, P.O. Box 1507, Morro Bay, CA 93443.

MORRO BAY ACCESS: *Mitchell Dr. and Doris Ave., Cuesta-by-the-Sea.* A path between bay-front homes affords a view of the south end of the Morro Bay Estuary. A parking space is reserved for wheelchair access; other visitors should park on the street.

MORRO DUNES NATURAL PRESERVE: *W. of Morro Bay Estuary.* A four-mile-long dune barrier peninsula separates the Morro Bay Estuary from the Pacific Ocean. Sand dunes, up to 100 feet high, are swept here by the wind and formed from sedimentary, igneous, and alluvial materials. The Morro Bay sandspit and dunes are a natural preserve within Montaña de Oro State Park. From the waterfront on Embarcadero Rd. in the town of Morro Bay, intrepid surfers sometimes paddle across the estuary, or a small boat will take you to the sandspit. Vehicles are not allowed on the sandspit, but pedestrian access is available from its south end; park in the first lot after you enter Montaña de Oro State Park on Pecho Valley Rd., and hike north. Look for California sagebrush, sand verbena, and coast fiddleneck.

Morro Dunes Natural Preserve

Estero
Bay

Morro Bay
Sandspit

*Morro
Bay
Estuary*

Baywood
Park

Montaña
de Oro
SP

Cuesta-
by-the-Sea

Los
Osos

PECHO RD

MONARCH LN

RD

VALLEY

LOS

OSOS

VALLEY

BAY BLVD

SOUTH

LOS

OSOS CT

RD

Monarch
Lane

Los Osos Oaks
State Reserve

PACIFIC

Hazard Canyon

PECHO

Montaña de Oro
State Park

Dune Tr

VALLEY

Spooner's
Cove

Islay

Cr

Bluff

Oats

Peak Tr

Valencia Peak Tr

Valencia Peak

Tr

Tr

Pt. Buchon

Rattlesnake

Coon

Creek

Flats Tr

Tr

Tr

Coon

Cr

OCEAN

N

0   0.5   1 Mile

0   0.5   1 Kilometer

See Legend on page 18 regarding
Marine Protected Areas offshore.

*Diablo
Cove*

Diablo Canyon
Power Plant

Morro Bay and Montaña de Oro State Park

	Sandy Beach	Rocky Shore	Trail	Visitor Center	Campground	Wildlife Viewing	Fishing or Boating	Facilities for Disabled	Food and Drink	Restrooms	Parking	Fee
Monarch Lane			•		•							
Los Osos Oaks State Reserve			•		•						•	
Montaña de Oro State Park	•	•	•	•	•	•		•		•	•	•

MONARCH LANE: *Off Pecho Rd., Cuesta-by-the-Sea.* A eucalyptus grove that harbors over-wintering monarch butterflies is located at the west end of Monarch Ln. Informal paths lead beyond the grove to sand dunes that are part of Montaña de Oro State Park.

LOS OSOS OAKS STATE RESERVE: *Los Osos Valley Rd., .6 mi. E. of South Bay Blvd., Los Osos.* This unit of the state park system offers a close look at several types of coastal habitat, including coast live oak woodland and coastal sage scrub. No facilities, but trails wind through the lichen-draped oak trees, which are up to 800 years old. Some of the trees are dwarfed from growing on the ancient sand dune. In spring, wildflowers grow in sunny spots, and visitors anytime may spot an oak titmouse, a brush rabbit, or a nest built by dusky-footed wood rats. Stay on the trail, as poison oak is prevalent, too. Dogs not permitted. Call: 805-772-7434.

MONTAÑA DE ORO STATE PARK: *S. end of Pecho Valley Rd., Los Osos.* This state park, one of California's largest, includes seven miles of coastline. Developed facilities are relatively few, but there is a wide array of visitor attractions. Golden yellow wildflowers give the park its name, and the colorful display from spring into summer is one of the park's most appealing features. Poison oak seems to grow everywhere here, however, so keep to established trails. Some 50 miles of trails are available to hikers, and some trails are open to equestrians and mountain bikers; obey posted restrictions. The park's beaches include long dune-backed stretches of sand and small cove beaches divided by rocky promontories.

Three-quarters of a mile inside the park entrance on Pecho Valley Rd., a spur road leads west to a parking lot and trailhead for the northernmost beaches of Montaña de Oro State Park. Hike north here to reach the miles-long stretch of sand on the Morro Bay sandspit. Back on the main road and heading south into Montaña de Oro State Park, trails take off at intervals, leading both inland, to destinations such as Valencia Peak, and to the shoreline.

The coast at Montaña de Oro State Park is made up of alternating shales and mudstones of the Monterey Formation. These rocks, some 10 to 15 million years old, were laid down in a deep marine basin, and they contain the remains of microscopic marine plants called diatoms that are rich in silica. During burial, the silica was mobilized, and it helped to cement the sediment into hard rock. Beds that originally contained many diatoms became well cemented and hard, while those that were poorer in diatoms remained softer and relatively less cemented. After these rocks were uplifted, tilted, and exposed at the surface, the less cemented beds proved more susceptible to wave attack, and they eroded into elongate bays, coves, and caves. The harder beds stand out as resistant points and fingers, especially prominent at low tide.

Montaña de Oro State Park

The main activity center of the park is at Spooner's Cove, where a historic dairy farmhouse now serves as the park ranger headquarters and visitor center, operated by the Central Coast Natural History Association; open Thursday through Sunday. In the canyon behind the visitor center are 50 campsites suitable for tents, trailers, or RVs up to 24 feet in length; tables and stoves are available and restrooms are nearby, but there are no showers or dump station. Spooner's Cove has a broad, sandy, easily accessible beach, flanked by sea cliffs of the Monterey Formation. Steeply dipping shale is exposed along the shoreline here, in planes that are flat, but not at all level.

From Spooner's Cove, take the trail north along the bluff for a view over the brilliantly white sandstone shore bordered by active surf and the deep blue ocean. To the south of Spooner's Cove, a blufftop trail overlooks a series of rocky fingers pointing southwest, with sandy pocket beaches between them.

A three-mile-long trail extending south of Montaña de Oro State Park, onto land owned by Pacific Gas and Electric Company, is planned; for information, call the California Department of Parks and Recreation: 805-528-0513.

Other camping facilities at Montaña de Oro include seven walk-in environmental campsites, in two groups, reached by trails that start one-tenth and seven-tenths of a mile short of the southern dead end of Pecho Valley Rd. Near the state park's entrance, the Hazard Canyon horse camp has two group campsites and three individual campsites with corrals and horse trailer parking; from Pecho Valley Rd., turn east just after entering the park. No horse rental available at the site, but rentals are available outside the park. All campsites at Montaña de Oro State Park, including the environmental sites, equestrian sites, and those at Spooner's Cove, can be reserved in advance; call: 1-800-444-7275.

One of the most striking sights of the Central Coast is one that few visitors will ever see. Set like a monument on a shelf above the ocean, the massive edifice and towering domes of the nuclear power plant at Diablo Canyon stand in stark contrast to the tumult of the sea and contours of the surrounding hillsides. Surrounded by 12,000 acres of undeveloped ranchlands owned by the plant's operator, the Pacific Gas and Electric Company (PG&E), the plant's dramatic facade is visible to the public only at a distance from the air or sea.

A mile-wide offshore security zone restricts access from sea, and a patrolled exclusion zone prevents access from land for all but employees and authorized visitors. Although glimpses of the site may be had from a new trail to be constructed on PG&E land south of Montaña de Oro State Park, topography precludes views from other public areas, including the Pecho Coast Trail.

The Diablo Canyon Power Plant is one of the state's largest energy generators, providing power for more than 1.6 million homes in northern and central California. About 1,400 employees operate the plant, which contains two pressurized water reactors in the twin 18-story domes and turbine generators capable of producing 2,200 megawatts of power in a hall the length of an aircraft carrier. Approximately 2.5 billion gallons of seawater are cycled daily through the plant as a coolant and released into Diablo Cove at a temperature approximately 21 degrees Fahrenheit warmer than surrounding waters. This "once-through" cooling system results in impacts to marine life due to thermal changes and direct entrainment of microscopic organisms.

Construction of the plant began in 1968, but commercial operation was delayed until the mid-1980s due to legal and regulatory challenges. The plant continues to generate controversy concerning seismic safety, security issues, impacts on the marine environment, and the plant's production and storage of nuclear waste. The plant is licensed to operate until 2025.

Due to security concerns, public tours of the plant have been discontinued. The Diablo Canyon Power Plant Information Center, located at 6588 Ontario Rd. in San Luis Obispo, is open to the public on weekdays, from 9 AM to 1 PM.

Diablo Canyon Power Plant

Coast Live Oak Woodland

C ALIFORNIA is home to twenty species of oaks, nine of which take the form of trees. Oak trees are found from one end of the state to the other, near the coast and in the interior. Oak forests, characterized by densely spaced trees that create deep shade beneath, are found in areas where moisture is plentiful, such as along streams. Oak savannas, in which trees are scattered widely across grassland, are found typically in much drier areas. A third type of oak tree–dominated plant community, oak woodland, is midway in its structure between forest and savanna. In the oak woodland, the canopies of trees are spaced far enough apart to allow sunlight to fall between the leaves, supporting an understory of shrubs and herbaceous plants and providing habitats for birds, mammals, insects, and other creatures.

Coast live oak are found from Mendocino County south into Baja California, in a strip 50 to 60 miles wide near the coast, although not on the immediate shoreline. The trees can also be found on Santa Cruz and Santa Rosa Islands. Coast live oaks are adapted to the coastal strip's moderate temperatures, which rarely reach the extreme highs and lows found farther inland, and to fog and moist breezes. In these ways, the coast live oak is unlike California's other oaks, some of which occur in very hot and dry inland valleys. Coast live oak woodland includes hollyleaf cherry, California bay laurel, coffeeberry, poison oak, and many other plants, in addition to the sturdy, often massive oak trees that dominate the structure of the woodland. Bats and many species of birds, including acorn woodpeckers, scrub jays, and oak titmice, can be found in the woodlands, along with black-tailed deer, gray foxes, ground squirrels, and jackrabbits. East of Morro Bay, at the El Moro Elfin Forest, stunted coast live oaks grow on ancient sand dunes. A limited supply of water and nutrients causes the normally shaped trees to grow to a height of only about ten feet.

Los Osos Oaks State Reserve

Coast live oak (*Quercus agrifolia*) is an impressive evergreen tree with a broad, dense crown and widely spreading branches. The dark green, holly-like leaves have distinctive fuzz on the underside. The genus name *Quercus* comes from the Celtic words *quer*, meaning "fine," and *cuez*, meaning "tree"; it truly is a fine tree. Unfortunately, a new disease organism, *Phytophthora*, that threatens oaks and other plant species arrived in California in the 1990s. *Phytophthora* spreads a disease known as Sudden Oak Death, and, once infected, an oak may die rather quickly. Native Americans revere the oak tree as a symbol of fertility and strength, and acorns were an important staple food for them.

Coast live oak

Many oaks have a delicate covering of gray-green **lace lichen** (*Ramalina menziesii*) draping their limbs. Lace lichen, sometimes referred to as Spanish moss, hangs down in curtains of long, branching strands that can reach five feet in length. Lace lichen is an epiphyte, meaning that although it must grow on a host plant, it does not hurt the host. Instead, it collects its nutrients from precipitation and wind. Like all lichens, lace lichen is actually a combination of a fungus and an alga. Lichens are formed when a fungus envelops either green algae or cyanobacteria. This combination allows the lichen to provide its own food source through photosynthesis. One of the unique qualities of lichens is that they are very sensitive to atmospheric pollution. Like blotting paper, lichens sop up contaminants in rainwater and fog but have no way to excrete them.

Lace lichen

The **oak titmouse** (*Baeolophus inornatus*) is the common, plain gray titmouse of the West. The species name *inornatus*, meaning "unadorned," is a good name for it, since the bird does not have a trace of contrasting colors. Its gray coat blends well with the trunks and branches of the oaks among which it forages. It is a charming bird, nonetheless, with its jaunty crest, sprightly manners, and its melodious whistle. Its call, *tschick-a-dee*, can be mistaken easily for the call of a chickadee. However, only a few bird species have the distinct crest of the titmouse. The oak titmouse is a secondary cavity nester, meaning it uses cavities created by other animals such as woodpeckers. The nests are usually in pine or oak trees, often in dead or dying wood.

Oak titmouse

Western scrub jay

The harsh call, aggressive manner, long blue tail, and blue head make the **western scrub jay** (*Aphelocoma californica*) easy to recognize. Scrub jays are a noisy lot, making the loud harsh *shreeeeeenk* and *wenk wenk wenk wenk* sounds that announce your arrival into oak woodland habitat. The western scrub jay is an inadvertent tree farmer. During the fall when acorns are ripe, scrub jays keep themselves busy by collecting the acorns and packing them into the ground. The jays bury many more acorns than they consume. In springtime the acorns germinate and may come up by the hundreds, helping to regenerate oak woodland. Next time you take a stroll through coast live oak woodland, look for newly sprouted oaks and then listen for the brazen calls of the scrub jay.

California quail

A quick movement across a forest path and a sudden burst of whirring wings is often the first indication of the presence of the **California quail** (*Callipepla californica*). The forward-curving plume rising from its crown distinguishes this state bird of California. Quail are loquacious little birds with a very distinct call. Listen for the scolding three-note call: *chi-ca-go*. The California quail is a ground-dwelling bird often seen scratching for seeds and insects in woodland areas. Look for the male as he makes his warning calls from an elevated perch while the female feeds the young or sits on the nest nearby.

Acorn woodpecker

The **acorn woodpecker** (*Melanerpes formicivorus*) is a black-and-white, clown-faced bird with a red crown and white eyes. The call of the acorn woodpecker sounds like laughter; listen for the nasal *waka waka waka* call resonating through the oaks. This species is highly social and usually lives year-round in social units, breeding in communal groups featuring one or two lead males, a harem of females (mostly sisters), and youngsters from the previous year. It is one of the most unusual breeding systems. Acorn woodpeckers are larder hoarders. Breeding groups gather acorns; the birds drill holes in a dead tree and stuff acorns into them. Studies have shown that the use of these granaries is one of the main reasons why acorn woodpeckers live in such large families. Only a large group can collect so many acorns and also defend them against other groups. The acorns are stored as winter provisions.

California sister butterfly (*Adelpha bredowii californica*) is a strikingly beautiful butterfly with a bright orange patch near the tip of the forewing. It is named for its black-and-white coloration on the forewing that resembles a nun's habit. The California sister is commonly found among oak groves, where it may be seen gliding among the higher branches of the trees. The females lay their eggs singly on oak leaves (*Quercus* spp.), which later serve the caterpillars as food. California sisters are usually bivoltine, meaning they have two generations per year. The caterpillars of the second generation enter diapause and sleep through the winter. In the spring, when the weather turns warm, the caterpillars complete metamorphosis. Look for the first adults in April as they hunt for mates and ideal egg-laying locations.

California sister butterfly

The **western fence lizard** (*Sceloporus occidentalis*) enjoys sitting on a prominent point, like a fence post or rock, where it can sun itself and watch for prey and predators. The bright blue patches along the sides of the body of the male give this lizard the common name of "blue-belly." The females lack this decorative coloring. The male fence lizard is territorial. He will fight with other males when they enter his territory. You may spy a male fence lizard doing pushups; this is how he shows his blue belly to other males in an attempt to warn them away.

Western fence lizard

Black-tailed deer (*Odocoileus hemionus*) occur along the coast of California and are distinguished by their black tails. During the rut, or breeding season, the bucks have terrific battles in which they knock their antlers against each other. During these fights, each buck tries to force the other's head down. If the antlers become locked together, both bucks will starve to death. The stronger, more virile bucks attract females and attempt to defend them against other bucks. Antlers are shed after the breeding season, and new antler growth begins immediately. Antlers are a true bone that grows covered with "velvet," a soft, skin-like tissue that carries nourishment and calcium to the antlers. After the antler growth is completed, the blood-supplying velvet is no longer needed, and it begins to fall or get rubbed off.

Black-tailed deer

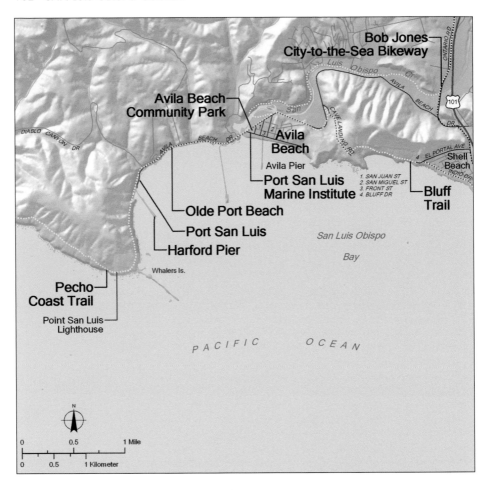

Olde Port Beach

Avila Beach

	Sandy Beach	Rocky Shore	Trail	Visitor Center	Campground	Wildlife Viewing	Fishing or Boating	Facilities for Disabled	Food and Drink	Restrooms	Parking	Fee
Pecho Coast Trail			•									
Harford Pier						•	•	•	•	•	•	
Port San Luis	•				•	•	•	•	•	•	•	•
Olde Port Beach	•					•	•			•	•	
Port San Luis Marine Institute				•								
Avila Beach Community Park								•		•	•	
Avila Beach	•	•				•	•		•	•	•	
Bob Jones City-to-the-Sea Bikeway			•									
Bluff Trail			•								•	

PECHO COAST TRAIL: *Diablo Canyon Dr. at Avila Beach Dr., N.W. of Port San Luis.* Docents lead hikes on the Pecho Coast Trail, starting from the Diablo Canyon Power Plant entrance. Hikes go to the Point San Luis Lighthouse, a three-and-a-half-mile roundtrip, or occasionally beyond the lighthouse. Reservations essential; call: 805-541-8735.

HARFORD PIER: *End of Avila Beach Dr., Port San Luis.* This 1,340-foot-long drive-on pier offers public fishing, fresh-off-the-boat fish sales, restaurants, and views of the bay. From the pier, halibut, salmon, mackerel, smelt, and sardines are caught in spring, and rockfish, sharks, and barracuda in summer and fall. No license required for fishing from the pier, which is lighted at night. Charter boat trips, fishing equipment, and fishing licenses available. Loudly barking sea lions gather under the pier.

PORT SAN LUIS: *End of Avila Beach Dr., W. of Avila Beach.* Port San Luis was constructed in the 19th century to serve the inland town of San Luis Obispo. A steam railroad was built between the town and the bay named San Luis Obispo, eventually connecting to the wharf built by John Harford in 1867. Steamships carrying passengers and freight docked at the pier regularly in the early days. The original wharf was destroyed by a storm in 1878. The port, originally known as Port Harford, served a thriving whaling industry in the late 19th century, and during World War II became a major West Coast oil port. Today, Port San Luis is a sport and commercial fishing center.

The harbor has approximately 200 private moorings and 35 seasonal guest moorings for boats up to 85 feet in length; a water taxi is available. Facilities include a 1,000-pound coin-operated boat hoist, diesel fuel and ice sales, a pumpout facility, boat wash-down area, trailered boat storage, and trailer parking. A boatyard offers a 50-ton hoist. Restrooms and coin-operated shower available. Overnight RV parking is allowed along the bluff overlooking the beach, east of Harford Pier. Fee for camping; see signs for payment instructions. Campsites are first-come, first-served; no information by phone about site availability. For information on Port San Luis Harbor District, see: www.portsanluis.com.

OLDE PORT BEACH: *Off Avila Beach Dr., Port San Luis.* It is permissible to drive down the slope onto this wide, sandy beach to load or unload boats; no parking on the sand. Four-wheel-drive vehicles recommended. Swimming, surfing, windsurfing, kayaking, and diving are popular. Restrooms and wheelchair-accessible ramp available; fire rings are on the beach in summer. Dogs permitted on leash. Call: 805-595-5400.

The small town of Avila Beach enjoys a favorable location on San Luis Obispo Bay. The beach faces south, and ocean waters are sheltered by Point San Luis, making this an attractive location for swimming and sunning. The protected bay also makes this a natural port. The Union Oil Company, founded in 1890 in the Ventura County town of Santa Paula, began operations in Avila in 1906. Petroleum storage tanks were constructed on the hill east of today's little town, and a pier was built farther west, connected by pipelines beneath Front St. Crude oil from the Santa Maria River Basin; from the Guadalupe Oil Field, which produced petroleum from the 1940s until 1994; and even from the San Joaquin Valley was stored in the tanks, and then transferred to tankers at the pier. The pipeline under Front St. was also used to transfer gasoline and diesel fuel brought in by ships from refineries elsewhere for storage in the tank farm and subsequent distribution.

Beginning in 1989, soil testing revealed that petroleum hydrocarbons had been leaking for years into the soil and groundwater under the town of Avila Beach. California's Regional Water Quality Control Board ordered the cleanup of an estimated 420,000 gallons of petroleum hydrocarbons. As Unocal Corporation began to remove

the leaked petroleum hydrocarbons in the 1990s, various restoration techniques were considered. One option was "solidification," a process in which concrete is injected into the ground, to lock petroleum hydrocarbons in place. Another is called "biosparging," in which air is pumped underground to vaporize the petroleum, allowing it to be pumped out. In the end, the main cleanup method was perhaps the most dramatic: removal of contaminated earth and sand, and replacement with clean fill material. About 11 percent of the town was removed, beginning in 1998, including demolition of most of the business district and part of the residential area, so that soil could be excavated for safe disposal. Part of the Avila Beach Pier was removed temporarily, and three acres of beach sand were replaced with clean sand.

The cleanup took place under an agreement between Unocal and local and state agencies and entities. Unocal funded the construction of a new community park at the west end of Front St. and of new Port San Luis Marine Institute facilities. Property owners have been rebuilding structures, and the low-key beach resort town is now concentrating on its role as a residential and visitor-serving, rather than industrial, community.

Avila Beach and Pier

PORT SAN LUIS MARINE INSTITUTE: *Avila Beach Community Park, Avila Beach.* This non-profit organization provides elementary and high school students with ocean-related educational opportunities and the chance to be a "marine biologist for a day." Programs take place onboard boats in Avila Bay and at a newly opened center that features marine exhibits. Call: 805-595-7280.

AVILA BEACH COMMUNITY PARK: *W. end of Front St., Avila Beach.* A pirate-ship play structure, basketball courts, picnic tables, and barbecue grills are set in a grassy park. Playground and picnic areas are wheelchair accessible.

AVILA BEACH: *S. of Front St., Avila Beach.* The wide sandy beach sheltered by Point San Luis is popular for swimming, kayaking, and water-related sports. Volleyball nets, picnic tables, barbecues, and outdoor showers are available, along with seasonal lifeguards and beach equipment rentals. A motorized beach wheelchair can be reserved for weekend use; call: 805-748-2434. No dogs allowed on Avila Beach from 10 AM to 5 PM.

The 1,635-foot Avila Pier is located in the middle of the beach. No license required for pier fishing. Bait and tackle are sold on the pier, which has a fish-cleaning station and public boat landing.

BOB JONES CITY-TO-THE-SEA BIKEWAY: *From Ontario Rd. parking lot to Avila Beach.* This bikeway is planned to link San Luis Obispo with the coast at Avila Beach. An

Patriot Sportfishing offers ocean fishing and whale-watching trips, 805-595-7200.

Central Coast Kayaks operates seasonally at Avila State Beach and Olde Port Beach, 805-773-3500.

The Sea Barn rents bodyboards, wetsuits, and other equipment across Front St. from Avila Beach.

Avila Beach Golf Resort, 805-595-4000.

existing mile-and-a-half-long portion of the paved path follows San Luis Obispo Creek through wooded areas and along a golf course, separated from vehicle traffic. Bicycling, skating, and walking are permitted on the path.

BLUFF TRAIL: *Between Cave Landing Rd., Avila Beach, and Bluff Dr., Shell Beach.* From a parking lot at the west end of Bluff Dr. off El Portal Ave. a paved path runs along part of the bluff, offering views of the Pacific Ocean. In the dense vegetation look for hummingbirds, wrentits, or a brush rabbit. Completion of the paved trail west to Cave Landing Rd. and Avila Beach is planned, forming a segment in the statewide California Coastal Trail for use by hikers and bicyclists.

Avila Beach

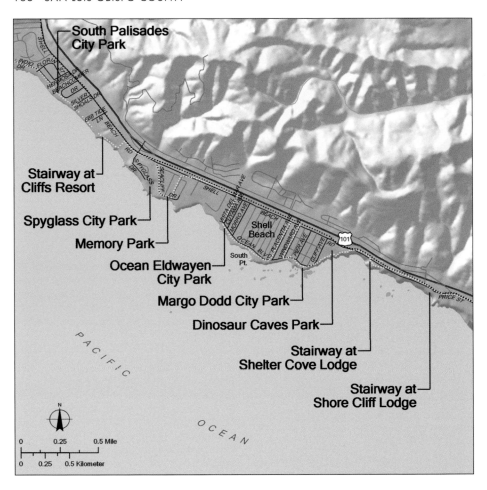

South Palisades
City Park

Stairway at
Cliffs Resort

Spyglass City Park

Memory Park

Ocean Eldwayen
City Park

Margo Dodd City Park

Dinosaur Caves Park

Stairway at
Shelter Cove Lodge

Stairway at
Shore Cliff Lodge

Shell
Beach

South
Pt.

PACIFIC

OCEAN

N

0   0.25   0.5 Mile

0   0.25   0.5 Kilometer

Margo Dodd City Park

Shell Beach

	Sandy Beach	Rocky Shore	Trail	Visitor Center	Campground	Wildlife Viewing	Fishing or Boating	Facilities for Disabled	Food and Drink	Restrooms	Parking	Fee
South Palisades City Park		•	•			•					•	
Stairway at Cliffs Resort	•	•	•			•					•	
Spyglass City Park		•								•	•	
Memory Park		•										
Ocean Eldwayen City Park	•	•				•					•	
Margo Dodd City Park	•	•				•					•	
Dinosaur Caves Park			•			•	•			•	•	
Stairway at Shelter Cove Lodge	•	•	•				•			•	•	
Stairway at Shore Cliff Lodge	•	•	•									

SOUTH PALISADES CITY PARK: *W. of Shell Beach Rd., N. of Silver Shoals Dr., Shell Beach.* South Palisades City Park offers benches and picnic facilities, along with great views of the ocean, surf, and seabirds. The park has several distinct parts, located along the high bluff at the north end of the Shell Beach community, which is part of the city of Pismo Beach. The city is working on creating a continuous shoreline park in the area.

From Shell Beach Rd., turn on Hermosa Dr. to Florin St., past a lawn with picnic tables and barbecue grills, surrounded by a residential neighborhood. At the south end of Indio Dr. off Florin St. is a path along the ocean bluff that is part of South Palisades City Park. The park can also be reached via Beachcomber Dr.; there is a parking lot inland of Shell Beach Rd., opposite the south end of the Beachcomber Dr. loop. A third approach to South Palisades City Park is at the end of Silver Shoals Dr., where there are picnic tables and barbecue grills. To the north and south of Silver Shoals Dr., surfers scramble down the bluffs along unimproved paths to reach the waves below.

Farther south along the ocean bluff is a small city park at the end of Ebb Tide Ln., where a short paved path runs along the bluff but does not yet connect to other parks. Future plans are in place to extend the blufftop path from Ebb Tide Ln. south to connect with the beach stairway at the Cliffs Resort.

STAIRWAY AT CLIFFS RESORT: *2757 Shell Beach Rd., Shell Beach.* A public path and stairway to the sandy beach are located at the north (upcoast) end of the Cliffs Resort hotel property. Signs mark the public beach parking area, along the northern margin of the main hotel parking lot. Additional beach parking is available on the inland side of Shell Beach Rd., opposite the Cliffs Resort. The beach path leads along a ravine vegetated with twisted coast live oak trees and willows, ending at a stairway that descends to the beach. A long curve of sand extends north along the base of the bluffs; this is a popular surfing spot. To the south of the beach access stairway is a rocky reef. In summer, look for willets and sanderlings feeding along the shore, while brown pelicans cruise by. In the thicket along the ravine, you might see a scrub jay, an oak titmouse, or a California towhee.

The Cliffs Resort also includes a public path, running along the blufftop on the seaward side of the hotel. The blufftop path connects to a new path located seaward of the new Dolphin Bay resort development. Parking for the Dolphin Bay blufftop path is at the south end of the development, off Shell Beach Rd.

Ocean Eldwayen City Park

SPYGLASS CITY PARK: *S. end of Spyglass Dr., Shell Beach.* This grassy park has picnic tables, barbecue grills, playground equipment, and parking. The bluff offers fine views of the ocean, and there is also access to the beach. A packed earth path leads down the highly eroded bluff to the shoreline, which is characterized by a series of rocky outcroppings, tidepools, and small pocket beaches. Surfers enjoy the waves here, although, depending on the tides, the ride may end at the base of the rocky bluff, rather than on a sandy beach.

MEMORY PARK: *Along seaward edge of Seacliff Dr., Shell Beach.* A blufftop lawn with benches, picnic tables, and nice views of waves breaking on the rocks below. No beach access or other facilities; street parking.

OCEAN ELDWAYEN CITY PARK: *Along Ocean Blvd., S. of Vista del Mar Ave., Shell Beach.* Ocean Eldwayen City Park offers vistas of sea and rocky shore; offshore seastacks are summer roosting places for cormorants and pelicans. There is a small blufftop lawn and benches and picnic tables in the park, as well as two stairways leading down to the beach. One beach stairway is a large concrete structure located near the end of Vista del Mar Ave. The second beach stairway is harder to see from the street; it is located along Ocean Blvd. between Cuyama and Morro Avenues.

The bluffs are eroded here, and the blufftop park is narrow. Depending on the tides, the beach below the park is sometimes narrow, too. At low tide, a dramatic reef of distorted beds of shale and sandstone of the Monterey Formation can be seen. The folded and contorted beds, which can also be seen in the sea cliff, are the result of compression and uplift that continues to affect much of the region. Street parking.

MARGO DODD CITY PARK: *Ocean Blvd. from Windward Ave. to Cliff Ave., Shell Beach.* At the end of Pier Ave., a stairway leads from this narrow blufftop park down to the rocky shore and scattered tidepools. Caves and tunnels in the rocks draw ocean kayakers. Street parking; no facilities.

At the downcoast end of Margo Dodd Park is a flat-topped islet, located only some 20 yards from shore, and looking as if it had merely floated away one day from the mainland. Harbor seals haul out on the little island, and seabirds such as brown pelicans roost here, offering an unusually close view from shore of wildlife. Pigeon guillemots nest on the cliffs.

DINOSAUR CAVES PARK: *Seaward of Price St. and Cliff Ave., Shell Beach.* This newly refurbished and spacious park features

Spyglass City Park

gently sloped wheelchair-accessible paths, grassy areas, restrooms with running water, benches overlooking the ocean and offshore sea stacks, and plenty of parking. There are no dinosaurs and no caves, other than sea caves that are not accessible from the park; the name stems from a one-time tourist attraction on the site. There is also no beach access from the park, but the views from the bluff are extraordinary.

STAIRWAY AT SHELTER COVE LODGE: *2651 Price St., Pismo Beach.* Park at the south (downcoast) end of the hotel, where there are restrooms and a landscaped path leading along the bluff edge. The paved trail is wheelchair accessible. There are nice views of rock formations and the ocean below. At the north end of the blufftop path, a stair descends to the shore; the last ten feet of stairs provide uncertain footing, but there is a solid railing to grasp. The narrow, 600-foot-long beach of mixed sand and pebbles faces a sea tunnel, where the surf thunders against the rocks.

STAIRWAY AT SHORE CLIFF LODGE: *2555 Price St., Pismo Beach.* A staircase structure at the north end of the hotel complex provides a way down the high bluff. From Price St., the stairway is half-hidden; look on the up-coast side of the restaurant building, where the stair is tucked behind the kitchen entrance. A small and sandy pocket beach lies at the base of the 120-foot-high bluff.

Monarch butterflies in eucalyptus foliage

Pismo Beach to Oceano

	Sandy Beach	Rocky Shore	Trail	Visitor Center	Campground	Wildlife Viewing	Fishing or Boating	Facilities for Disabled	Food and Drink	Restrooms	Parking	Fee
City of Pismo Beach	●	●	●	●	●	●	●	●	●	●	●	
Pismo Beach Pier							●	●	●	●	●	
Pismo Coast Village RV Resort					●			●	●	●	●	●
Pismo Lake Ecological Reserve			●			●						
Pismo State Beach North Beach Campground	●		●		●	●		●		●	●	●
Butterfly Trees						●					●	
Pismo State Beach	●		●	●	●	●	●	●	●	●	●	●
Sand and Surf RV Park					●			●	●	●	●	
Pismo State Beach Oceano Campground	●		●	●	●	●	●	●	●	●	●	
Oceano Memorial Campground					●		●	●	●	●	●	
Oceano Community Park							●	●		●	●	

CITY OF PISMO BEACH: *Hwy. 101, 11 mi. S. of San Luis Obispo.* Travelers enjoy sweeping views of beach and sand dunes at Pismo Beach, one of the few spots north of Gaviota where the ocean is visible from Hwy. 101. Pismo Beach retains the feel of an old-time beach resort. The original wharf, constructed in 1881 near the site of the present Pismo Beach Pier, was built for shipping out farm produce and bringing in lumber, but tourism soon gained in importance. From 1895 until the 1920s there was a dance pavilion at the foot of the pier. During the summer months the pavilion was surrounded by a tent city; tents could be rented for eight dollars a week. Hotels, a skating rink, a bowling alley, and other attractions drew visitors, some of whom arrived on the Southern Pacific Railroad, which began serving the area in the 1890s.

The name "pismo" is derived from *pismu,* the Chumash word for the asphaltum tar that seeps through natural fissures in the ground and sea floor of central California. The Chumash people used the tar to caulk their plank canoes and to make baskets watertight.

PISMO BEACH PIER: *Near end of Pomeroy Ave., Pismo Beach.* Although not quite as big as the original pier, which was more than a quarter-mile long, the Pismo Beach Pier offers plenty of space for strolling and viewing the ocean. Large decks extend out to the sides of the pier, providing space for fishing for barred surfperch, flounder, or sanddabs. The pier is open 24 hours and is lit at night; it is also wheelchair accessible. Concession

Recreational outfitters include:

Horseback riding on the beach: Pacific Dunes Riding Ranch, 805-489-8100.

Kitesurfing and paragliding equipment: Xtreme Big Air, 805-773-9200.

Beach Cycle Rentals, near the pier, 805-773-9400.

Central Coast Kayaks, 805-773-3500.

Surf shops include:

Moondoggies, 805-773-1995.

Pancho's, 805-773-7100.

Pismo Beach Surf Shop, 805-773-0134.

stand, bait sales, and fishing equipment rentals are on the pier, and shops and restaurants are nearby. Restrooms, beach showers, play equipment, and parking are located at the foot of the pier; additional parking is at the end of Addie St., near Pismo Creek.

Pismo Beach is a popular surfing spot, although no surfing is allowed within 500 feet of the pier. Beach volleyball nets, available for public use on a first-come, first-served basis, are located on the sand north of the pier, near the end of Wadsworth Ave. The pier and nearby beach are part of Pismo State Beach, but are maintained by the city of Pismo Beach; call: 805-773-7039.

PISMO COAST VILLAGE RV RESORT: *165 S. Dolliver St. (Hwy. One), Pismo Beach.* A very large and well-equipped facility, the Pismo Coast Village RV Resort includes 400 RV sites with full hookups, located adjacent to Pismo State Beach. Facilities include a general store, restaurant, clubhouse, swimming pool, recreational facilities, and laundry. Wheelchair-accessible facilities. Fee for camping. For reservations, call: 805-773-1811.

PISMO LAKE ECOLOGICAL RESERVE: *N. 4th St. and Five Cities Dr., Pismo Beach.* The reserve is located just west of the 4th St. exit off Hwy. 101. The reserve encompasses 30-acre Pismo Lake, which was formerly much larger but is still a good birding site. Shorebirds using the marsh include overwintering sandpipers, plovers, and yellowlegs. Herons, egrets, and rails are permanent marsh residents, and waterfowl such as grebes, dabbling ducks, and diving ducks visit the marsh during their migration. Managed by the city of Pismo Beach; call: 805-773-7039.

PISMO STATE BEACH NORTH BEACH CAMPGROUND: *S. Dolliver St., .6 mi. S. of Pomeroy Ave., Pismo Beach.* This campground has 103 campsites, each with stove and table, but

Pismo State Beach

without RV hookups. Pay showers available. An RV dump station is located on Le Sage Dr. off Hwy. One, one-tenth mile north of Grand Ave. The beach is nearby, seaward of the sand dunes. A path and boardwalk leads south from the campground through the dunes. Reservations are accepted for camping during the summer months; call: 1-800-444-7275. For park information, call: 805-489-2684 or 805-473-7220.

BUTTERFLY TREES: *Pismo State Beach North Beach Campground, Pismo Beach*. Pismo State Beach harbors the largest colony of overwintering monarch butterflies in the U.S. Volunteer docents are available daily from 10 AM to 4 PM during the prime butterfly months of November through February. Docent talks are given daily at 11 AM and 2 PM. A temporary gift shop is set up at the grove during the butterfly season. Park along S. Dolliver St. south of the entrance to Pismo State Beach North Beach Campground.

PISMO STATE BEACH: *Runs from Wilmar Ave. in Pismo Beach to S. of Arroyo Grande Creek.* This large state park includes miles of beaches and dunes, extending from the town of Pismo Beach south through Grover Beach and Oceano to the Oceano Dunes State Vehicular Recreation Area. In Pismo Beach, there is public access to the wide, sandy beach from numerous street ends, from Wilmar Ave. in the north, past the Pismo Beach Pier, to Addie St.

Pismo State Beach offers two campgrounds, the North Beach Campground and the Oceano Campground. Also within the park is a nine-hole, three-par golf course with putting and pitching green, located at 25 W. Grand Ave. in Grover Beach. Open daily from sunrise to sunset; affordable greens fees. Call: 805-481-5215.

SAND AND SURF RV PARK: *1001 Pacific Blvd. (Hwy. One), Oceano.* This large campground offers tent campsites and RV sites with hookups. There are 232 campsites with picnic tables and barbecues, as well as a heated swimming pool, hot showers, laundry, and basketball and volleyball courts. Pets welcome; some facilities are wheelchair acces-

sible. For reservations, call: 1-800-330-2504; for information, call: 805-489-2384.

PISMO STATE BEACH OCEANO CAMPGROUND: *555 Pier Ave., Oceano.* Located between the beach and Oceano Lagoon is a campground, open year-round, with 82 sites. Forty-two sites have RV hookups; restrooms with hot showers available. The ocean is a short hike away. The mile-long Guiton Trail leads around the lagoon through a willow thicket, where birders may spot many migrating warblers in spring and fall. For camping reservations, call: 1-800-444-7275. For park information, call: 805-489-2684 or 805-473-7220.

The Pismo Nature Center next to the campground entrance contains exhibits about monarch butterflies, resources of the lagoon, and the history of the Chumash indigenous people. There is also a small shop offering books and gifts. Volunteer-maintained interpretive gardens contain labeled native plants from the dunes and other parts of San Luis Obispo County. From June 1 to Labor Day, the Nature Center is open on Monday, Wednesday, Friday, and weekends from 1 to 4 PM; the remainder of the year, the center is open on Friday, Saturday, and Sunday only. Docent-led walks begin at the Nature Center. For information on programs offered by the Central Coast Natural History Association, call: 805-772-2694.

OCEANO MEMORIAL CAMPGROUND: *Air Park Dr. at Mendel Dr., Oceano.* This San Luis Obispo County park offers 22 campsites, located near Pismo State Beach. All sites have electricity, water, and sewer hookups; tents are also welcome. Restrooms and showers available. Campsites are available on a first-come, first-served basis, and the campground fills up quickly on weekends. Open year-round; call: 805-781-5930.

OCEANO COMMUNITY PARK: *Norswing Dr. at Mendel Dr., Oceano.* This park includes picnic facilities, a children's playground, basketball court, playfields, and restrooms. A group picnic area can be reserved; call: 805-781-5930. Across Norswing Dr. is an arm of Oceano Lagoon, surrounded by lawns.

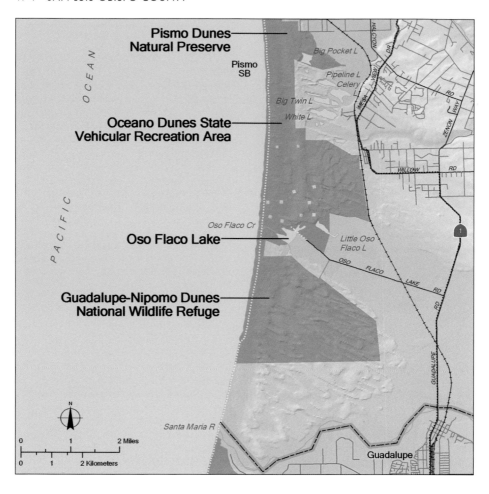

Pismo Dunes Natural Preserve

Oceano Dunes State Vehicular Recreation Area

Oso Flaco Lake

Guadalupe-Nipomo Dunes National Wildlife Refuge

Surf fishing

Pismo State Beach to Nipomo Dunes

	Sandy Beach	Rocky Shore	Trail	Visitor Center	Campground	Wildlife Viewing	Fishing or Boating	Facilities for Disabled	Food and Drink	Restrooms	Parking	Fee
Pismo Dunes Natural Preserve			•			•						
Oceano Dunes State Vehicular Recreation Area	•			•		•	•		•	•	•	
Oso Flaco Lake	•	•				•	•	•	•	•	•	
Guadalupe-Nipomo Dunes National Wildlife Refuge	•		•			•	•					

PISMO DUNES NATURAL PRESERVE: *Inland of the beach, S. of Arroyo Grande Creek, Oceano.* The Preserve extends about a mile and a half south of Arroyo Grande Creek, and three-fourths of a mile inland from the beach. Arroyo Grande Creek is located about one mile south of Pier Ave. Hiking trails traverse the Preserve, where sand dunes, freshwater lakes, and riparian habitats support a large number of rare endemic plants. Enter the Pismo Dunes Natural Preserve on foot, from the beach; the area is off-limits to vehicles. Information about plants and animals found in the area is available at the Pismo Nature Center at the Pismo State Beach Oceano Campground. Call: 805-473-7220.

OCEANO DUNES STATE VEHICULAR RECREATION AREA: *W. of Hwy. One, S. of Oceano.* This unique park offers recreational opportunities not found elsewhere on the California coast. The Oceano Dunes State Vehicular Recreation Area is the only state park unit where vehicles can be driven on the beach. Over five miles of beach are open to vehicle use, and 1,500 acres of dunes are open to off-highway motor vehicles.

Vehicles enter the beach at Grand Ave. in Grover Beach or Pier Ave. in Oceano. One mile south of Pier Ave. is the beginning of the off-highway vehicle use area, marked by a post on the beach bearing the number 2. Motorcycles, dune buggies, and other off-highway vehicles can be used on the beach or in the dunes south of Post 2, except where fences or signs indicate areas closed to vehicular use. Off-highway vehicles must bear a current registration sticker from the California Department of Motor Vehicles; registration fees support the operation of the Oceano Dunes State Vehicular Recreation Area. Suitable vehicles can be rented in the Oceano area. Off-highway vehicle drivers should take part in safety training, which is mandatory for operators under the age of 18. Helmets are required.

Camping is permitted in the Vehicular Recreation Area south of Post 2. Bring your own water and carry out your trash; vault toilets are provided. An RV dump station is located on Le Sage Dr., off Hwy. One north of Grand Ave. Although the camping area is open year-round, vehicle access to it can be made difficult by winter storms or blowing sand. Four-wheel-drive vehicles are recommended at all times to reach the camping area. Camping reservations are also recommended; call: 1-800-444-7275. Fees are charged. Popular recreational activities on the beach include surfing, fishing, horseback riding, and swimming. Lifeguards are on duty from June through Labor Day. Dogs must be leashed at all times. Beach wheelchairs can be checked out at the beach entrances at Pier and Grand Avenues.

Gathering Pismo clams is a long-standing activity. In the 19th century, the clams were so numerous that they were taken by plowing the beach, and farmers used the clams for animal feed. As the population of Pismo clams declined, commercial harvesting ended, and now the bag limit for recreational clamming is ten Pismo clams. Clammers must have a fishing license and an accurate measuring tool in order to take only Pismo clams that are at least four and half inches long; smaller clams must be reburied immediately. For information, call: 805-473-7220.

OSO FLACO LAKE: *W. end of Oso Flaco Lake Rd., 3 mi. W. of Hwy One.* From the entry kiosk to this state park unit, a trail leads through riparian woodland to Oso Flaco Lake. A boardwalk continues across the water to the broad, sandy beach. Look for American bitterns along the margin of the lake. To the beach and back is about two and a half miles. Open 8 AM to sunset; no dogs or bicycles permitted. Fee for parking. A beach wheelchair for use on the Oso Flaco Lake Trail can be checked out at the Dunes Center in the nearby town of Guadalupe at 1055 Guadalupe St. (Hwy. One). Call ahead to reserve the wheelchair: 805-343-2455. For information on Oso Flaco Lake, call: 805-473-7220.

GUADALUPE-NIPOMO DUNES NATIONAL WILDLIFE REFUGE: *S. of Oceano Dunes State Vehicular Recreation Area, W. of Hwy. One.* This relatively new wildlife refuge includes four square miles of the Guadalupe-Nipomo Dunes, with two linear miles of beach. The Guadalupe-Nipomo Dunes National Wildlife Refuge was created to protect a relatively intact portion of the dune complex and the breeding habitat of several imperiled species, including the western snowy plover and California red-legged frog. Also found here are the western spadefoot toad, horned lizard, and legless lizard, along with black bear, deer, mountain lions, and bobcats. Many species of wintering waterfowl frequent the refuge, and spring brings a colorful wildflower show to the dunes, highlighted by giant coreopsis, dunedelion, and La Graciosa thistle. There are few facilities for visitors, but ample resources for wildlife-watchers and beach-strollers. Fishing for surfperch is possible along the shore; state fishing regulations apply.

Enter the Guadalupe-Nipomo Dunes National Wildlife Refuge from Oso Flaco Lake; walk west along the boardwalk to the beach, then continue south one mile to the refuge. The Hidden Willow Valley Trail leads from the beach through foredunes and back dunes; return the way you came. Keep on the trail to protect sensitive dune plants. Trail signing is limited, but improvements are planned. From March 1 through September 30 each year, access to the upper beach and foredunes is closed to protect the breeding area of the western snowy plover. For information, call: 805-343-9151.

Oso Flaco Lake

Between 1946 and 1994, over 200 oil wells were in production in the Guadalupe Dunes. Peak production in the 2,700-acre Guadalupe Oil Field reached 4,500 barrels per day. Because the area's crude oil was typically very thick and heavy, a kerosene-like petroleum product called diluent was added to assist in transportation of the crude oil. In 1988, surfers noticed petroleum hydrocarbons on the shore near the mouth of the Santa Maria River. It soon became apparent that over a period of years diluent had been inadvertently released into the ground from pipelines and storage tanks. Between 8.5 and 20 million gallons of spilled petroleum hydrocarbons were estimated to be present in the soil or in a layer up to six feet thick, floating over groundwater, comprising about 90 separate diluent plumes. For comparison, the *Exxon Valdez* oil spill, the largest in American history, deposited about 11 million gallons of petroleum into Alaska's Prince William Sound in 1989.

Unocal Corporation, the operator, began a shutdown of the field, and cleanup efforts commenced in 1990. Cleanup methods included extraction wells, excavation of beach sand, and construction of temporary walls in the sand to keep the diluent from flowing into the river and ocean. These efforts were complicated by natural meandering of the mouth of the Santa Maria River and by the fact that the dunes and river provide habitat for rare and endangered animals and plants, including the western snowy plover, tidewater goby, and beach spectacle pod. Initial efforts to excavate, treat, and return contaminated soil to its original location were abandoned after treated soil continued to show toxic characteristics. Instead, plans were revised to include trucking treated, non-hazardous soil to a landfill site elsewhere in the county.

The spill resulted in criminal charges and fines against Unocal Corporation. As partial response for the spill, the company has funded public education efforts, public beach access improvements, and restoration of dune habitat. Restoration efforts include the propagation and planting of rare plants such as La Graciosa thistle and the removal of invasive exotic plants.

Santa Maria River mouth

Guadalupe-Nipomo Dunes

THE SANTA MARIA RIVER flood plain and its terraces to the north and south host perhaps the most extensive dune system in coastal California. The dunes are so extensive here because the two things that are needed to grow dunes—sand and wind—are in abundance. The Santa Maria River provides a steady sand supply, and the prevailing northwesterly winds blowing off San Luis Obispo Bay provide the means to move it inland.

Four miles wide and over twelve miles long, the Nipomo Dune complex comprises the Callender Dunes, the Guadalupe Dunes, and the Mussel Rock Dunes. From overhead, individual dunes are shaped like parabolas, with long, parallel tails pointing toward the coast, from which prevailing winds blow. As the tails become stabilized by vegetation, sand is blown out of the center of the dune, and the dune advances landward. The longest of these parabolic dunes have tails over a mile long.

The Nipomo Dunes formed during several different episodes, separated by long periods of little activity. The various episodes of active dune formation are related to fluctuations in sea level. The most inland dunes presumably grew when sea level was higher than today, perhaps during the interglacial period about 75,000 to 140,000 years ago. During the last ice age, sea level was more than 400 feet lower than at present, and the shoreline in this area was many miles seaward of its present location. Dunes may have formed along the shoreline, but were submerged and eroded away when sea level started rising again about 18,000 years ago.

The youngest dunes are those forming today, of sand blown off Pismo Beach. These young dunes are moving actively, and few plants have established a toehold on them. They extend, at most, two miles inland from the beach. The young dunes are encroaching on a series of much older dunes that have been stabilized by vegetation. The type of vegetation may indicate age—oak chaparral is established on the oldest dunes, whereas sparse grasses and bushes stabilize younger dunes. In addition, on the older dunes soils are red from iron oxide. As the youngest dunes have encroached on the older dune fields, they have blocked off depressions, creating natural dams to hold back the area's many lakes such as Celery, White, and Oso Flaco Lakes. The advancing dunes have also displaced Nipomo and Los Berros Creeks, and the creeks now flow around the dune lobe before joining the Santa Maria River and Arroyo Grande, respectively.

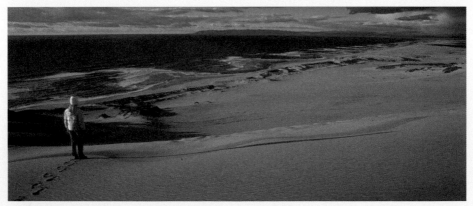

Guadalupe-Nipomo Dunes

The **globose dune beetle** (*Coelus globosus*) inhabits California's coastal sand dune habitats. It is most commonly found beneath dune vegetation in foredunes and sandy hummocks. The dune beetle leaves a distinct track on the beach that resembles a labyrinth; however, its footprints cannot be seen. The dune beetle burrows beneath the surface of the sand, leaving a collapsed tunnel behind. Adult dune beetles cannot fly, however this species has been documented at scattered dune localities from Mendocino County to Ensenada in Baja California.

Globose dune beetle

Pholisma (*Pholisma arenarium*), also known as sand food, is a strange non-photosynthetic root parasite. The plant produces a peculiar mushroom-shaped cluster of purplish flowers. Pholisma prefers sandy soil; it occurs along the central and southern coast of California in dune habitat, but it can also be found in the Mojave and Sonoran deserts. It has a fleshy, scaly, subterranean stem and an unusual growth habit. The entire plant lives below the surface of the sand, with only the flower head pushing above sand during early spring. The scales on the stems are actually modified leaves. The scaly stem gets all of its vital nutrients from nearby host plants. Pholisma sends out "pilot roots" two feet below the surface of the sand. When they reach the vicinity of a host shrub, the pilot roots send out special roots that connect and penetrate the host root. The connection allows pholisma to absorb carbohydrates and amino acids from the photosynthetic host shrub.

Pholisma

Giant coreopsis (*Coreopsis gigantea*) is a peculiar shrub that looks like something from a Dr. Seuss book. It grows to about six feet tall, with a main trunk up to five inches thick, giving the coreopsis the appearance of a small tree. The genus name *Coreopsis*, "bug-like," comes from the Greek words *koris*, "bug," and *-opsis*, indicating a resemblance. This is in reference to the fact that the seeds look a lot like ticks. This interesting plant has spectacularly bright yellow flowers that bloom between March and April. During the rest of the year, coreopsis looks like nothing more than a brittle, thick brown stump. Giant coreopsis is found on rocky cliffs and exposed slopes and dunes along the coast from San Luis Obispo to Los Angeles County.

Giant coreopsis

Beach spectacle pod

Beach spectacle pod (*Dithyrea maritima*) is a low-growing, small, perennial herb in the mustard family. The flowers are white and cross-shaped. The seed pods have two sections, each surrounded by a rim, which gives the appearance of a pair of spectacles. Beach spectacle pod is limited to coastal foredunes and active sand and dune scrub. It once ranged from San Luis Obispo to Baja California. However, currently it is known to occur only in the dunes of San Luis Obispo and Santa Barbara Counties and on San Nicholas and San Miguel Islands. It is usually found approximately 50 to 300 meters from the surf in dunes where the sand is relatively unstable. Beach spectacle pod is protected pursuant to the California Endangered Species Act as a threatened species.

Dunedelion

Dunedelion (*Malacothrix incana*), a member of the sunflower family, is a compact, somewhat fleshy perennial herb with sticky white sap. Its yellow, dandelion-like flower heads and its dune habitat together give it the common name, dunedelion. Dunedelion is found sporadically in the foredunes and coastal scrub areas along the immediate coast from Ventura County to San Luis Obispo County. It is found also on San Nicolas, Santa Rosa, and San Miguel Islands. Dunedelion is identified by the California Native Plant Society as a plant of limited distribution.

La Graciosa thistle

La Graciosa thistle (*Cirsium loncholeppis*) is a spiny member of the sunflower family with purple and white flowers. The bushy purplish flower heads occur in wide, tight clusters at the tips of the stems. La Graciosa thistle occurs adjacent to coastal dune slack ponds (a type of wetland that occurs in wind-carved depressions between dunes), freshwater ponds, and in brackish marsh habitat at the mouth of the Santa Maria River. All extant populations of this species are located in San Luis Obispo and Santa Barbara Counties. La Graciosa thistle is protected by federal and state endangered species laws due to threats that include groundwater pumping, oil field development and remediation, and competition from aggressive native and non-native plants. There are only 17 known populations.

Page opposite: Santa Barbara Point, Santa Barbara County

Santa Barbara County

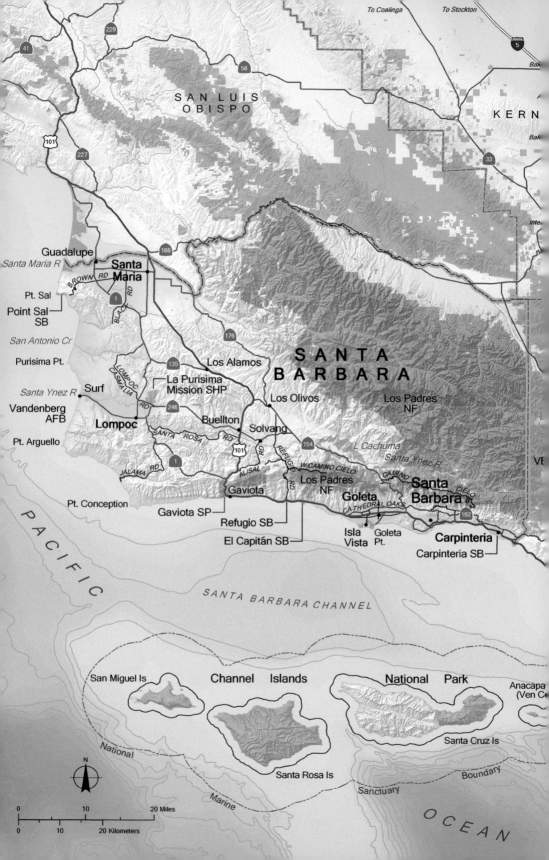

Santa Barbara County

SANTA BARBARA County is situated where northern California meets southern California. Offshore to the south of the county lies the Southern California Bight, the ocean within the arc of coastline from Point Conception to the Mexican border. The sea floor within the bight is like a plain, where the ocean over the continental shelf is relatively shallow but punctuated by peaks and canyons. The abrupt turn in the coast at Point Conception, compared to the orientation of the coastline north of the point, creates a gyre, or eddy, in the Santa Barbara Channel. The result of this gyre is the Southern California Countercurrent, a shallow current that flows generally from east to west and bathes the coast south of Point Conception with water about two degrees warmer than that north of the point.

Santa Barbara County's southern coast is sheltered moderately from wave energy by the four Northern Channel Islands, and the south-facing orientation of the beaches increases solar heat gain. North of Point Conception, the coastal climate is often cooler and windier. Santa Barbara County offers a panoply of activities: swimming, parasailing, surfing, bodyboarding, sailing, kayaking, diving, fishing, spear fishing, wildlife viewing from shore or boats, and photography. Sunning and beach strolling are equally popular. Not far from the shoreline is the Los Padres National Forest, which offers mountain scenery, wilderness trails, and more.

Most of Santa Barbara County's population lives near the county's south coast, where coastal parks feature varied facilities, such as fully equipped marinas, volleyball nets and tennis courts, and beach restaurants that offer sunset dining. West of Goleta, by contrast, facilities and services are more limited. State parks from Gaviota to El Capitán State Beach include a total of 11 miles of shoreline, not all of it contiguous, but only a handful of coastal access points. Visitors heading to beach parks between Lompoc and Goleta should stock up on gasoline and supplies before setting out; a convenience store at Jalama Beach County Park provides food service year-round, and small stores operate seasonally at Gaviota State Park, Refugio State Beach, and El Capitán State Beach. Despite its urban amenities, Santa Barbara County is still mainly rural; agriculture remains the county's largest industry, with gross annual production valued at nearly one billion dollars.

The Santa Barbara Metropolitan Transit District serves a number of beach areas and parks in the communities of Goleta, Isla Vista, Santa Barbara, Montecito, Summerland, and Carpinteria. Visit the Transit Center at 1020 Chapala St. in Santa Barbara. Call: 805-683-3702 or see www.sbmtd.gov.

For visitor information, contact:

Guadalupe Chamber of Commerce, 805-343-2236.

Lompoc Valley Chamber of Commerce, 805-736-4567.

Goleta Chamber of Commerce, 805-967-4618.

Santa Barbara Region Chamber of Commerce, 805-965-3023.

Carpinteria Chamber of Commerce, 805-684-5479.

In Santa Barbara, there is a visitor information center at the corner of Garden St. and Cabrillo Blvd.

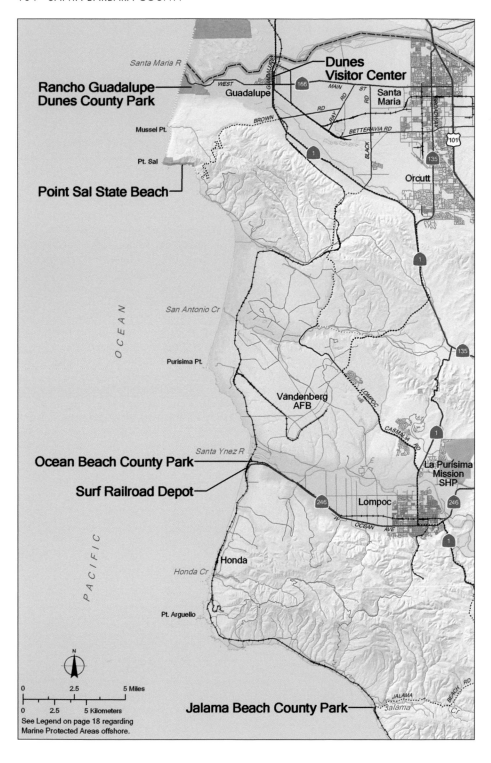

Santa Maria R

Dunes Visitor Center

Rancho Guadalupe Dunes County Park

WEST
Guadalupe
166
MAIN
RD
ST
Santa Maria

BROWN
RD
RAY
RD
BLACK

BETTERAVIA RD

101

Mussel Pt.

1

135

Pt. Sal
Orcutt

Point Sal State Beach

OCEAN

San Antonio Cr

Purisima Pt.

135

Vandenberg AFB

LOMPOC

CASMALIA RD

1

Santa Ynez R

Ocean Beach County Park
La Purísima Mission SHP

Surf Railroad Depot

246
Lompoc
246

W OCEAN AVE

1

PACIFIC

Honda

Honda Cr

Pt. Arguello

N

0   2.5   5 Miles
0   2.5   5 Kilometers
See Legend on page 18 regarding
Marine Protected Areas offshore.

JALAMA
BEACH RD
Cr

Jalama Beach County Park
Jalama

Santa Maria River to Jalama Beach

	Sandy Beach	Rocky Shore	Trail	Visitor Center	Campground	Wildlife Viewing	Fishing or Boating	Facilities for Disabled	Food and Drink	Restrooms	Parking	Fee
Dunes Visitor Center				•		•				•	•	
Rancho Guadalupe Dunes County Park	•					•	•	•		•	•	
Point Sal State Beach	•	•	•			•						
Ocean Beach County Park	•					•	•	•		•	•	
Surf Railroad Depot	•						•	•		•	•	
Jalama Beach County Park	•				•	•	•	•		•	•	•

DUNES VISITOR CENTER: *1055 Guadalupe St., Guadalupe.* The visitor center on Hwy. One offers information about the large dune complex located north and south of Guadalupe. The nonprofit Dunes Center offers guided walks to the Guadalupe-Nipomo Dunes National Wildlife Refuge, Oso Flaco Lake Natural Area, and the Rancho Guadalupe Dunes County Park. Some hikes highlight geology, birds of the dunes, photography, or stargazing. Educational programs are offered for schoolchildren and researchers. Open 10 AM to 4 PM, Tuesday through Sunday; call: 805-343-2455.

RANCHO GUADALUPE DUNES COUNTY PARK: *End of Main St., 5 mi. W. of Guadalupe.* A massive dune complex extends north and south of the mouth of the Santa Maria River. The only public road access to the dunes south of the river is at Rancho Guadalupe Dunes County Park, which offers paved parking, picnic tables, and restrooms. The dunes are open to exploration from approximately October 1 through March 1. During the remainder of the year, when western snowy plovers are nesting here, only the beach is open to public access, and temporary fences block access to dune areas. Dogs are not permitted in the park, even in vehicles, during the snowy plover nesting season. Ocean fishing is popular all year; barred surfperch and strawberry perch are taken frequently. The park is managed by the nonprofit Center for Natural Lands Management; open sunrise to sunset. Call: 805-343-2354. To the south of the park on private land, the dunes near Mussel Point reach an elevation of over 500 feet, the highest beach dunes in the western United States.

POINT SAL STATE BEACH: *W. of Hwy. One, S.W. of Guadalupe.* A secluded beach is located nine miles west of Hwy. One. Landslides have closed Brown Rd. to vehicles at a point four miles from Hwy. One; the Santa Barbara County Parks Department is studying options for a substitute beach trail nearby. The park remains open to those willing to hike in; no facilities. Call: 805-733-3713.

OCEAN BEACH COUNTY PARK: *End of Ocean Park Rd., 9.2 mi. W. of Lompoc.* From Lompoc, take W. Ocean Ave. (Hwy. 246) eight miles to Ocean Park Rd.; turn north to a 28-acre park located next to the mouth of the Santa Ynez River. There is a picnic area with barbecue grills, sheltered from prevailing winds by the railroad berm. Restrooms have running water. The park, a favored location for birders, overlooks the estuary, salt and freshwater marshes, and mudflats. The wetland is an important resting and foraging site for migrating shorebirds and waterfowl and provides habitat for raptors such as the northern harrier and the osprey. There is a nesting colony of California least terns at the river mouth. Several species of amphibians, including the foothill yellow-legged frog and the tiger salamander, live in the freshwater lagoon. The County Parks Department plans to construct a boardwalk along the estuary.

Ocean Beach County Park provides seasonal access to the broad sandy beach, part of Vandenberg Air Force Base. The ocean beach is prime nesting habitat for the endangered western snowy plover, and from March 1 through the end of September, public beach access may be closed entirely. Please respect all posted restrictions. During the remainder of the year, public access to the Air Force Base's beach is available from 1.1 miles to the north of the Santa Ynez River to 3.5 miles south of the river. The park may also be closed prior to and during rocket launches. Ocean Beach County Park is open 8 AM to sunset; call: 805-934-6123.

SURF RAILROAD DEPOT: *End of W. Ocean Ave., 9.5 mi. W. of Lompoc.* An operating railroad station provides parking, restrooms, and access to a Vandenberg Air Force Base beach. Beach access is across the railroad tracks, equipped with warning lights and bells; use caution when crossing. Public beach access during the snowy plover nesting season from March 1 through the end of September is available only from the mouth of the Santa Ynez River south for a distance of one-half mile. Please respect all posted restrictions; violations may lead to closure of the entire beach during the nesting season. The beach may also be closed prior to and during rocket launches. Call: 805-606-1921.

JALAMA BEACH COUNTY PARK: *Jalama Beach Rd., 14.2 mi. W. of Hwy. One, S. of Lompoc.* At the terminus of Jalama Beach Rd., this is the most remote of California's public beaches along Hwy. One. The park offers day use and camping in an open, relatively unsheltered setting. There are 112 campsites for tents or RVs; no reservations taken. Facilities include picnic tables with barbecue grills, group picnic areas, a playground, and a snack stand and store open daily, year-round. There are hot showers, including a wheelchair-accessible shower.

Activities on the broad sandy beach include strolling, surf fishing, swimming, surfing, and windsurfing. Lifeguard service available during the summer. This is often a windy spot; surfing is best in the morning or when the wind is low. Beach wheelchair available from a lifeguard or park staff. The county park boundary is at Jalama Creek, where a small estuary draws songbirds to its riparian vegetation. Vandenberg Air Force Base allows one mile of beach access northwest of Jalama Creek. The Amtrak *Coast Starlight* railroad trains cross Jalama Creek on a high trestle. Fee for day use and camping. For information, call: 805-736-3504.

Jalama Beach County Park

Destroyer shipwrecked, 1923

Following Fleet Week activities in 1923, thirty U.S. warships left San Francisco on September 8 for home base in San Diego. Among them were the 14 destroyers of Squadron 11, each 314 feet long with a displacement of 1,250 tons and 27,000-horsepower engines. Radio directional finder equipment that relied on shore radio transmitters was newly available but not universally relied upon. Navigators still used dead reckoning to infer their position from speed and distance traveled. Shipboard radar was yet to be invented.

Squadron 11 moved at a speed of 20 knots in three columns, in close formation; only 150 yards separated the ships of each column, front to back. Captain Edward H. Watson, commodore of the squadron, was in the lead ship, *Delphi*. As darkness fell and the convoy neared Point Arguello, a dense fog formed.

At 8:30 PM, the navigator aboard the *Delphi* calculated that the squadron was now south of Point Arguello, some 12 miles from shore. However, the onshore radio directional finder signal indicated instead that the ship was still north of the point and, moreover, close to land. The navigator suggested taking a sounding of the water's depth, but slowing the convoy would have spoiled the ships' schedule and formation.

Concluding that the way was clear to enter the Santa Barbara Channel north of San Miguel Island, the captain ordered a change of course to eastward, just before 9:00 PM. Speed was held at 20 knots. Within minutes, the *Delphi* scraped and then struck bottom with tremendous force, stopping the ship immediately and throwing its crew and contents to the decks. Captain Watson sent word to the other closely following ships, but there was no time to avoid more wrecks.

The destroyers *S. P. Lee, Nicholas, Woodbury, Young, Chauncey,* and *Fuller* followed one after the other, striking the shore or each other at a place known as Honda, located a few miles north of Point Arguello. The remaining seven destroyers were able to slow in time to avoid disaster.

The sailors abandoned ship, assisted by the crews of nearby fishing boats. On shore, a resident railroad employee heard the commotion and called for rescuers, including a doctor from Lompoc. More than half of the 800 officers and crew were rescued during the night and sent via train to hospitals; others were rescued the next day. The final tally of dead or missing was 23 men. The event is still the worst peacetime disaster suffered by the U.S. Navy.

Allain Manesson-Mallet, *Description de l'univers*, Tome V., *Nouveau Mexique et Californie*, Paris, 1683

The Island of California

B OTH the name and image of California began as the label for a fictional island paradise, created by a Spanish novelist in the early 16th century. About 1510, Garcí Ordoñez de Montalvo wrote the best-selling novella *Las Sergas de Es-plandián*, or *The Adventures of Esplandián*, a story about the exploits of Esplandián and his father during the medieval Crusades in the Holy Land. Montalvo described how the hero of his story battled troops of an "infidel" Amazon race, allied with the Turks and led by the beautiful queen Calafía, who wore golden armor and rode wild beasts. The author gave the name *California* to the island that he described as being located east of the Indies; possessed of great wealth, gold, and pearls; protected by rocky cliffs; and inhabited by Calafía's race of beautiful, black Amazon women. It was a place described as a rugged country, where men were captured during raids of vessels, used solely for breeding purposes, and then fed to trained griffins.

The Spanish reading public of that time found the image of Montalvo's mythical queendom irresistible, and their excitement was reinforced by the New World explorations of Juan Ponce de León, Hernándo de Soto, Francisco Vásquez de Coronado, and Hernán Cortés. Cortés actually found gold, silver, gems, and dark-skinned natives during his expedition of 1519 to present-day Mexico, prompting Spain to finance expeditions in search of El Dorado, Quivíra, and Cíbola, "northern" settlements described by the natives as having vast treasure.

In 1534 Cortés was sent to colonize the "island" to the northwest of Mexico that was already being called California. Such vain quests for riches and an easier route to the Orient produced new maps, but little else, and the colony near present-day La Paz foundered in the harsh climate. Cortés returned to Spain bankrupt and disgraced about the same time mariners of the day sailing up the east coast of Baja California as far as the Colorado River delta began to depict a mythical "Straits of Anían" on their charts. At least one version of the mysterious waterway had it connecting the Colorado River with the Pacific Ocean, thus making California an island. The Straits of Anían were sought by others who had an entirely different waterway in mind, the so-called Northwest Passage, connecting the North Atlantic Ocean and the North Pacific.

Juan Rodríguez Cabrillo was the first Spaniard to travel along the Pacific coast of today's state of California, seeking the Straits of Anían. Sailing north from Baja California in 1542, he was usually greeted in a friendly manner by the indigenous people and gave many coastal locations their first Spanish place names. He ventured as far north as Monterey Bay; there, winter weather forced a retreat to San Miguel Island, where he died due to an infected broken arm. His pilot, Bartolomé Ferrélo, continued north when the weather improved, reaching Oregon's Rogue River before turning back, having not found the straits he sought.

The island myth persisted as prospects for European colonization of California in the latter half of the 16th century failed to materialize. Spain under Charles V was at war in Europe and struggled to suppress political strife at home while holding its possessions abroad. A lucrative trade with the Spanish colonies in the Orient developed between Manila and Acapulco, but exploration of the California coast came to a halt, and the San Francisco estuary went undiscovered by Europeans for another two centuries. Pirates who attacked and plundered the Manila galleons became more familiar than the Spaniards with California's coastline.

A genuine opportunity to map more accurately what is now California came in 1602 with the voyage of Sebastián Vizcaíno, charged by the Spanish government with locating and mapping harbors suitable for settlements and support of crews returning from the Orient. Vizcaíno stopped at and named many places, claiming the land for Spain. He was most impressed with Monterey Bay, which he described as *un puerto famoso*, ideal for future settlement. However, despite the relatively friendly natives and Vizcaíno's improved charting, the voyage produced no Anían, no San Francisco estuary, and no obvious mineral wealth.

Another century passed before the California coast began to take a more scientifically correct form on world maps. After Jesuit Padre Eusebio Kino's extensive explorations and missionary work in the late 1600s, a turning point came in 1747, when Spain's King Ferdinand VII actually issued a formal decree that California was indeed part of the mainland. Juan Gaspar de Portolá's 1769 land expedition along the coast followed with the "discovery" of the San Francisco estuary, and in 1775, Juan Manuel de Ayala's ship *San Carlos* became the first Spanish vessel to enter the bay. Ayala's pilot, Jose de Canizares, spent 40 days surveying the bay and produced the first *Plano del Puerto de San Francisco* in 1776.

Despite the conclusive evidence of explorations before 1800, the depiction of California as an island on many world maps remained unchanged. It was as if a romantic attachment to the idea led cartographers of the time to resist redrawing California attached to the mainland. However, as ship navigators began to use precise chronometers for more accurate longitude calculations and land survey parties began to use trigonometric methods for mapping, the true shape of the coastline was confirmed. By the year 1900 federal mapping agencies such as the U.S. Coast Survey and the U.S. Geological Survey had produced highly detailed topographic and hydrographic maps of the entire west coast of North America with coastal features such as capes, bays, and offshore islands, all in their proper locations.

———————   • • •   ———————

It turns out, after all, that California *is* a kind of island. The coastal area that lies west of the San Andreas fault is a part of the Pacific Plate of the earth's crust, not the North American Plate, as is the rest of the continent. Many plant and animal species in California, particularly along the Central Coast, differ dramatically from those of the rest of North America. From the steep Santa Lucia Mountains of Big Sur in Monterey and San Luis Obispo Counties to the rolling hills of Santa Barbara County and the broad, flat Oxnard plain in Ventura County, the area boasts an unparalleled and astoundingly rich and diverse natural environment, unique in that it contains many northern *and* southern species adapted to exist at their latitudinal limits.

California boasts not only a unique physical geography, but also a remarkable cultural diversity. Before European settlement, California was occupied for around 13,000 years by the continent's most populous native peoples north of Mexico, estimated to number over 300,000 in 1535. The indigenous peoples of California spoke approximately 100 different languages. The cultural landscape included hunting and gathering sites, villages, trails, sacred sites and burial grounds, a homeland in which the people coexisted on a sustainable basis with their ecosystem.

European civilization began in California with a distinctly different form of relating to the surrounding physical environment. The 300-year period of Spanish and Mexican rule produced the *Californios*, a people with an independent and somewhat isolated society, noted for its cattle ranching economy and its class-based hospitality. Among contemporary English-speaking visitors, California's *gente de razon*, or "people of reason," acquired a reputation for romantic rituals and festivities, but were also responsible for the first wave of displacement of native peoples and their culture.

The initial period of American settlement continued the legacy of Hispanic California's fierce independence. Before California's admission to the United States in 1850, American Californians actually framed a constitution as an independent republic. Following the discovery of gold in 1848, thousands of immigrants from throughout the world converged on California, and almost overnight the economy and infrastructure to support mining operations sprang into being. A new multi-ethnic diversity resulted, reflecting the presence of not only gold seekers, but also refugees from revolutions in Europe and Asia. California's new residents did their best to keep their cultural ways intact, but at the same time they became part of a new way of life, socially, economically, and politically.

The dream of instant wealth and unlimited opportunity in California fed an era of speculative development and corruption that followed the gold rush and continued into the early 20th century. Advertising campaigns fueled a series of massive real estate booms that began in the 1880s and brought many immigrants, some with utopian ideas, who created cooperatives, communes, and experimental ventures in living and working. Not only wealthy capitalists, but also members of the middle class enjoyed the dream of the California good life. They created an image of California as a place where, with some good luck and hard work, instant wealth was achievable. Reflecting this image are the fruit crate labels of the time, which depict idealized scenes of beautiful produce, landscapes, and people, seeming to confirm California as the place to live a healthy, long, and successful life.

As Nobel and Pulitzer Prize–winner John Steinbeck illustrated in his books *The Harvest Gypsies* (1936), *In Dubious Battle* (1936), and *The Grapes of Wrath* (1939), this image masked the reality that existed for many, especially for immigrant workers and transplants from the drought-stricken Midwest during the worldwide depression of the 1930s. The Hollywood film industry, before and after World War II, not only generated great wealth, but also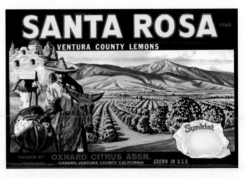
helped to expand and promote cultural fantasies about "making it" in California. Writers from Henry Miller to Robinson Jeffers to Jack Kerouac suggested visions of the landscape, and in particular California's Central Coast, that inspired and attracted yet another generation of immigrants. Perhaps the cultural island of California is both a symbolic object of hope for people the world over and, for its inhabitants, a sort of civilization all its own. California society remains connected to the land, its ecological treasures, and its unique history, with human and environmental diversity and a distinct style of expressing its character.

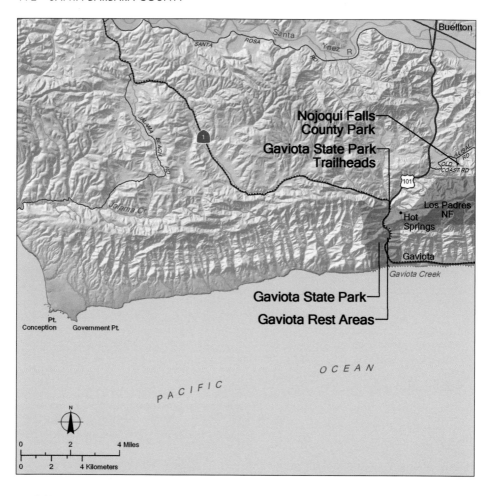

Point Conception Lighthouse, not accessible to the public

Point Conception to Gaviota

	Sandy Beach	Rocky Shore	Trail	Visitor Center	Campground	Wildlife Viewing	Fishing or Boating	Facilities for Disabled	Food and Drink	Restrooms	Parking	Fee
Gaviota State Park Trailheads			•		•				•	•	•	
Gaviota State Park	•	•	•		•	•	•	•	•	•	•	•
Gaviota Rest Areas								•	•	•		
Nojoqui Falls County Park			•			•			•	•	•	

GAVIOTA STATE PARK TRAILHEADS: *E. and W. of Hwy. 101, 3 mi. N. of Gaviota.* A trailhead on the east side of Hwy. 101 provides access to a hot spring and trails in the northeastern part of Gaviota State Park. Take the Hwy. One exit from Hwy. 101, and turn south to the parking lot. Hike up the fire road from the parking area and turn left at the first junction. About three-quarters of a mile from the parking area, where the fire road crosses a creek, turn uphill on a narrow trail to a series of spring-fed pools filled with warm, sulfurous, blue-hued water. The fire road continues upward, and hikers can proceed to Gaviota Peak, a distance of three miles one way, and other points in the Los Padres National Forest. On a clear day, the trails offer broad views of the Santa Barbara Channel and Channel Islands.

On the west side of Hwy. 101 at the Hwy. One interchange, head west on Hwy. One for one-half mile to San Julian Rd. and turn south to a parking area, restrooms, and trailhead. Trails lead south into Gaviota State Park; the beach is four miles distant.

GAVIOTA STATE PARK: *Hwy. 101 at Gaviota Beach Rd., Gaviota.* At Gaviota, the southbound traveler winds through the dramatic gorge of Gaviota Creek and emerges onto a narrow coastal terrace overlooking the Santa Barbara Channel. Gaviota Creek flowed to the coast more or less along its present course prior to the uplift of the Santa Ynez Mountains. As the mountains slowly rose due to compression associated with movement of the earth's crust, the river kept pace by cutting downward, maintaining its course, and carving the deep canyon we drive through today. Uplift also resulted in tilting of the sedimentary rocks lying on the south flank of the Santa Ynez range. Resistant rocks form dramatic outcroppings called flatirons, dipping toward the sea, along the flanks of the mountains. These dipping beds also can be seen in the sea cliffs along the Gaviota coast.

Gaviota Pier, Gaviota State Park

Gaviota State Park includes 2,776 acres of land and five miles of shoreline. Day-use parking and restrooms are located at the beach, near an 800-foot-long railroad trestle and the park's main entrance. There are 41 campsites that accommodate tents, RVs up to 27 feet long, and trailers up to 25 feet long. No hookups available; a dump station is located at El Capitán State Beach. No reservations taken for campsites. Between October 1 and March 31, the campground is open only from Friday through Sunday. Campground restrooms and showers are wheelchair accessible.

The fishing pier has a two-ton boat hoist. No fishing license is required for fishing from the pier; bait and tackle are available. Lifeguards are on duty during the summer. A beach wheelchair is available from the camp host. Fees for camping and day use. An additional day-use area, San Onofre Beach, is located off Hwy. 101, one-and-a-quarter miles east of the Mariposa Reina intersection. Watch for the state park sign at a dirt pull-out on the southbound side of the highway; use caution crossing the railroad tracks. For park information, call: 805-968-1033.

Explorer Juan Rodríguez Cabrillo stopped at what is now Gaviota while sailing along the coast in 1542. Juan Gaspar de Portolá and his crew traveled overland through the area in 1769 on their search for Monterey Bay. Padre Juan Crespi noted in his journal that the soldiers called the place La Gaviota, after the seagull they killed there. The beach at Gaviota once served as a harbor for Mission Santa Inés, located in the town of Solvang.

Wildlife in the park's grasslands includes western meadowlarks, turkey vultures, gray foxes, and coyotes. Areas of chaparral and coastal sage scrub are home to the California

Gaviota State Park

The ecological and climatic dividing line between north and south that Point Conception represents is apparent to sport anglers. Those fishing at Gaviota Pier or elsewhere east of Point Conception may encounter fish such as Pacific barracuda or yellowtail that are common to the open ocean waters off southern California and Mexico. Other species found in northern California waters, such as coho salmon, are taken only in small numbers south of Point Conception. From piers at Gaviota State Park, Goleta Beach, and Santa Barbara, anglers take tomcod, spotfin croaker, sand shark, and jacksmelt during the summer months; halibut fishing is good in spring and early summer. During the winter, pier fishing for several types of surfperch is good. The reefs and extensive kelp beds found along the south-facing shore of Santa Barbara County are good places for fishing for rockfish or bonito. Shore fishing along the Ventura County coast may produce walleye, kelp bass, or surfperch; rockfish may be caught from the rocky shore south of Point Mugu.

No fishing license is required for sport fishing off the Gaviota Pier or other public fishing piers in California's ocean waters. A public fishing pier is one that has unrestricted access to the general public. Some breakwaters and jetties qualify as public fishing piers, but check first, or simply obtain a fishing license. Two no-license-required free days are offered each year, usually in early and late summer; contact the Department of Fish and Game for upcoming dates.

Anyone 16 years old or older fishing in California from shore, a boat, or from a jetty or wharf that does not meet the definition of a "public pier" must have a fishing license displayed so that it is plainly visible above the waist. A fishing license is required to take not only fish, but also mollusks, invertebrates, or crustaceans. If a fishing license is required, an ocean enhancement stamp must also be obtained for ocean fishing south of Point Arguello in Santa Barbara County.

Whether or not a fishing license is required, limits apply on how many fish can be taken. The limits are different for sport fishing and commercial fishing, and the rules are complicated. In general, sport fishing rules provide that any one person may have possession of no more than 20 finfish of all species, and no more than 10 of any one species. For many species, additional limits apply. Certain gear restrictions also apply to fishing for rockfish, lingcod, and salmon, including from a public pier.

Before you go on a fishing trip, contact the Department of Fish and Game for their Ocean Sport Fishing Regulations, which are published annually. The booklet of regulations contains the rules for taking finfish, crustaceans, and other sport species, and also has maps, information on marine protected areas, and public health advisories. For even more current information or to check regulations while you are fishing, call: 831-649-2801. The department also publishes fish identification guides and fish bulletins, and the department's website contains images of many different sportfish species.

For more information, see the Department of Fish and Game's website at www.dfg.ca.gov, or contact a Marine Region office at Monterey, 831-649-2870; Morro Bay, 805-772-3011; or Santa Barbara, 805-568-1231.

Gaviota Pass

thrasher, white-crowned sparrow, scrub jay, and dusky-footed wood rat. Gaviota Creek has supported a small, sporadic steelhead spawning run. The perennial stream also provides habitat for the red-legged frog, and Pacific tree frogs may be found in the freshwater marshes. West of the pier, dipping layers of the Monterey Formation are eroded into intriguing forms.

GAVIOTA REST AREAS: *Hwy. 101, Gaviota Pass.* A small but immaculately tended and landscaped rest area is located on each side of Hwy. 101. The sandstone peaks of the Santa Ynez Mountains, part of Gaviota State Park, surround the rest stops, which feature interpretive panels about the area's biological and geological resources.

NOJOQUI FALLS COUNTY PARK: *Alisal Rd., off Hwy. 101, 5.5 mi. S. of Buellton.* This pleasant park features a year-round waterfall in a shady oak woodland in the Santa Ynez Mountains. Turn east off Hwy. 101 at Old Coast Rd., 4.7 miles south of Buellton, and follow Alisal Rd. to the park entrance. Nojoqui Falls County Park can also be reached from Solvang along Alisal Rd. There are family and group picnic areas among spacious lawns shaded by large trees, a playground, and ballfields; look for acorn woodpeckers tending their stash of acorns in holes in the tall palm trees.

A trail leads up the canyon to Nojoqui Falls, about a ten-minute walk from the parking area. A narrow cascade of water falls 160 feet over a shale and sandstone cliff into a small pool. Maidenhair ferns, common on the northern California coast but rare in Santa Barbara County, grow on the north-facing rocks above the pool. Open 8 AM to sunset; for information, call: 805-934-6123.

Even before San Francisco was joined to the U.S. east coast by railroad in 1869, a railroad line from San Francisco Bay south along the coast was contemplated by backers of the transcontinental route. An 1857 survey studied alternate routes around Point Conception and through Gaviota Pass, in Santa Barbara County. The Southern Pacific Railroad, formed in 1865, laid track that reached the Salinas Valley by the 1870s. A branch line to Monterey, and later Pacific Grove, commenced passenger service in 1880, the same year the railroad's backers opened the grand, original Hotel Del Monte in Monterey. Construction of the railroad line reached San Luis Obispo over the Cuesta Grade in 1894, then Guadalupe in 1895, and Surf in 1896. Meanwhile, railroad construction proceeded northward from Los Angeles, reaching Ventura and Santa Barbara in 1887. Building the roadbed on the Rincon coast required major earthmoving; for lack of an easily developed route, a plank road had been built on the beach to serve coastal wagon traffic. The bridging of numerous ravines complicated construction and delayed completion of the remaining 56 miles of track in Santa Barbara County, between Surf and Ellwood, for another 14 years. Until then, rail customers bound from San Francisco to Los Angeles were forced to transfer at Surf to a stagecoach for the bumpy ride to Santa Barbara, where they could board a train for their destination. Not until 1901 did the entire coast railroad route open for passenger service.

The railroad drew revenue not only from passengers, but also from oil fields along the route, as well as agricultural products from the Salinas valley and the Oxnard plain. *Sunset* magazine, started by the Southern Pacific Rail-road, promoted tourism to sites on the line, including the missions built during the Spanish period. Railroad stations, including the one in Santa Barbara that opened in 1906, and the rebuilt Hotel Del Monte that opened in 1926 in Monterey, featured the Mission Revival architectural style. As popular as the coastal rail journey was with tourists, the Southern Pacific reportedly drew more revenue from the canneries at Monterey than from visitors to the area. In 1923, twelve long-distance passenger trains were in daily operation, six each way, bearing names such as the *Seashore Express*, the *Sunset Limited*, and the *Lark*.

Steam power was still in use on coastal passenger trains until the mid-1950s. Autos and airplanes had taken most of the passengers by the time Amtrak took over most U.S. long-distance passenger service in 1971. Today, the *Coast Starlight* and *Pacific Surfliner* trains run daily along California's Central Coast. For about 107 miles between Surf and Ventura, the line follows the coastline closely. The area near Point Conception and Point Arguello, including the dramatic 37-mile stretch from Gaviota to Surf, is inaccessible to visitors other than those riding the train; much of the land traversed is part of Vandenberg Air Force Base or is private land, with no public beach access.

Southern Pacific *Daylight* near Goleta, 1947

Kelp

Giant Kelp Forest

THE GIANT KELP, *Macrocystis pyrifera*, is an alga that forms submarine forests in the cold Pacific waters of both North and South America. North American kelp forests are scattered along the coast from Alaska to Baja California but are particularly well developed within the Southern California Bight, where they generally grow on rocky substrates in a narrow band between water depths of about 30 and 60 feet. In sheltered areas, where wave action is less likely to dislodge the kelp from the substrate, the inner edge of the kelp forest sometimes creeps closer to shore, and in very clear water where enough light reaches the bottom for baby plants to thrive, the outer edge may descend into deeper water. Occasionally, a few giant kelp plants are even seen in the lower intertidal zone and in water over 120 feet deep.

Each plant is anchored to the bottom by a holdfast that gives rise to fronds with leaf-like blades buoyed up by gas-filled bulbs. Each frond lives for about six months, but new fronds are constantly being produced near the sea floor. Under optimal conditions, kelp fronds may grow from one to two feet per day, soon reaching the surface, where they form extensive floating canopies. In shallow areas, feather boa kelp may contribute to a mixed canopy. In California, the upper four feet of the canopy is periodically cut off and harvested to make alginate, which improves the texture and stability of a vast array of manufactured products ranging from cosmetics to ice cream. Although over 100,000 tons of kelp are harvested commercially most years, new fronds soon form a new canopy.

In the water, the intertwined fronds of large kelp plants rise to the sea surface in massive columns, giving a diver an impression similar to that received by a visitor to a forest of giant sequoias. As in the terrestrial forest, the understory is dark and open, with flecks of sunlight dancing overhead. Giant kelp forests are often compared to coral reefs and rain forests because of their great biological productivity and biodiversity. Perhaps a thousand other species find a home within a large kelp forest.

A giant kelp plant may occasionally live as long as a decade, but the average life span of an adult is only two or three years. Most mortality is associated with large waves. Dislodged plants tend to tangle with other plants, which are, in turn, torn from the substrate. Although they generally eat drift algae, sea urchins occasionally feed on and kill adult kelp. Lost adults are generally soon replaced. Large numbers of tiny kelp recruits appear every one to three years following periods when the influx of nutrient-rich cold water coincides with lots of light reaching the seafloor. Recruitment events often are preceded by storms that disturb the bottom and create bare space for colonization.

At one time, much of the giant kelp in southern California was found growing along the sandy Santa Barbara coast, which is usually sheltered from large waves by the Channel Islands. This was a particular type of giant kelp, perhaps a different species, whose anchoring holdfasts were low and spread broadly across the sandy sea bottom instead of attaching to submarine rocks. The large mass of the holdfast and its trapped sediments stabilized the kelp and kept it in place. However, large waves associated with the 1982–83 El Niño tore up these sand-based kelp forests and most have never returned. Like ferns, adult kelp produce spores that are dispersed in currents. The spores settle on the bottom and develop into microscopic individuals that reproduce to form another generation of large spore-producing adults. Today there are few spore sources, which may limit recolonization of the sand bottom. On the other hand, perhaps spores are available, but environmental conditions are not yet right.

Southern sea otter

The **southern sea otter** (*Enhydra lutris nereis*) is the largest member of the weasel family. Adult sea otters reach a size of about four feet long. Their hind feet are webbed, and they have the thickest fur in the animal kingdom. Sea otters dine on sea urchins, crabs, clams, and other species of marine invertebrates that occur in the kelp forest. When a sea otter comes to the water's surface with its meal, the otter lies on its back and uses its stomach as a table. Sometimes a sea otter uses a rock to help open the hard shells of its prey. Sea otters live in a narrow band of ocean along the coast and rarely venture very far from the shore. Sea otters were brought almost to extinction by unregulated hunting during the 1700s and 1800s. Due to protective measures that were instituted by the California state government and the federal government, southern sea otters are making a comeback. If you have time for a stroll along the bluffs in Point Lobos State Reserve, you are likely to spy one or two. Listen for the distinctive sound of the sea otter cracking its prey open with a rock.

Brown sea hare

The **brown sea hare** (*Aplysia californica*) is a type of gastropod. Gastropods are creatures such as snails and slugs. The brown sea hare has ruffled flaps along its back and two pairs of antennae. It is herbivorous and feeds on a variety of algae, primarily red and brown algae, which gives the animal its dark coloration. The brown sea hare has the ability to release a reddish-purple ink from glands under its mantle when threatened. Some researchers have proposed that the ink acts as a screen or decoy to deter predators. Brown sea hares are used extensively in laboratory studies of the neurobiology of learning and memory.

Decorator crab

Decorator crabs (*Loxorhynchus crispatus*) camouflage themselves with tiny seaweeds and animals like anemones, sponges, and bryozoans. The crabs attach algae and/or small marine invertebrates to various portions of their shell using hooked setae, which are Velcro-like bristles. This allows them to hide from predators, like cabezon, by blending into the surrounding rocks. Decorator crabs, like all crabs, have to cast off, or molt, their exoskeleton in order to grow. During the molting process, a decorator crab may save some of its living decorations and use them to decorate its new shell.

One of the largest sea stars in the world is the **sun-flower star** (*Pycnopodia helianthoides*). Adults can grow to over 30 inches in diameter. They usually have 20 to 24 arms. The sunflower star is a voracious predator. It preys on other invertebrates such as clams, sea cucumbers, sea urchins, and snails. It is relatively quick moving for a sea star, moving at speeds of up to 40 inches per minute. Once a sunflower star locates its prey, it attaches its many-tubed feet to the shell and pulls. When the prey finally tires, the sunflower star's stomach will extrude and devour the meat.

Sunflower star

The **California sheephead** (*Semicossyphus pulcher*) is a beautiful, long-lived reef fish. This species is a "protogynous hermaphrodite"; what this means is that all sheepheads begin their lives as females. When the fish are seven or eight years old, the ovaries become testes and the fish function as males for the rest of their lives. California sheepheads generally occur in rocky kelp areas near shore, in water from 20 to 100 feet deep. A sheephead hunts during the day; however, at night it moves to a crevice or cave and secretes a mucous cocoon around itself. Predators searching for prey are unable to detect the sheephead scent through the mucous cover.

California sheephead

The genus name for **feather boa kelp** (*Egregia menziesii*) comes from the Latin word *egregious*, meaning "remarkable." This unusual seaweed is known for its physical strength and its beauty. The feather boa kelp can be found attached to rocks in the intertidal and subtidal zone. Feather boa kelp is one of the largest intertidal brown kelps, and it can grow up to 60 feet long. The stipe, or stem, is strongly flattened, and covered with little bumps that feel like rough sandpaper. From either side of the stipe arise numerous crowded blades of various lengths. Some of the blades are swollen into floats. The many blades along the side of the stipe give this kelp the look of a feathery scarf. Look for feather boa kelp washed up on the beach after a big storm.

Feather boa kelp

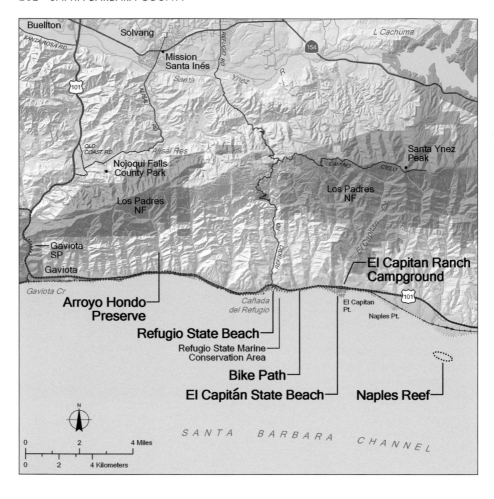

Refugio State Beach

Arroyo Hondo to Naples Reef

	Sandy Beach	Rocky Shore	Trail	Visitor Center	Campground	Wildlife Viewing	Fishing or Boating	Facilities for Disabled	Food and Drink	Restrooms	Parking	Fee
Arroyo Hondo Preserve			•	•		•		•		•	•	
Refugio State Beach	•	•	•		•	•	•	•	•	•	•	•
Bike Path			•									
El Capitán State Beach	•	•	•		•	•	•	•	•	•	•	•
El Capitan Ranch Campground				•	•	•		•	•	•	•	•
Naples Reef							•					

ARROYO HONDO PRESERVE: *Off Hwy. 101, 4 mi. W. of Refugio State Beach.* This limited-access, 782-acre preserve in the canyon of Arroyo Hondo contains features of natural and historical interest. The preserve is managed primarily for the use of school groups and research purposes; it is also open to the public on the first and third weekends of each month. Docents lead walks on trails along the sycamore-shaded creek and into areas of oak woodland and chaparral. From higher elevations, visitors may see sweeping vistas of the Santa Barbara Channel and the islands that bound it. Hikers may also explore on their own during public open times. A visitor center is located in the circa-1850 Ortega home, one of several in the area built by descendants of José Francisco de Ortega, a member of the Portolá expedition of 1769 and a founder of the Presidio of Santa Barbara in 1782. Restrooms and the path to the Ortega adobe are wheelchair accessible. The Arroyo Hondo Preserve is managed by the Land Trust for Santa Barbara County; for information and reservations, call: 805-567-1115. Donations in lieu of an entry fee are encouraged.

REFUGIO STATE BEACH: *S. of Hwy. 101 at Refugio Rd., 15 mi. W. of Goleta.* At the mouth of Cañada del Refugio is a popular state beach, offering day use and camping. The creek mouth forms a small lagoon, shaded by palm trees that lend a semi-tropical atmosphere to the setting. Day-use parking is located at beach level west of the creek, and there is a seasonal store that offers food and beverages, beach equipment rentals, fishing licenses, and bait. Picnic tables are sited under the palms. A beach wheelchair is available; call ahead to reserve it. This is a popular beach to launch a kayak; swimming is also good here, and lifeguards are on duty during the summer months. Day-use hours are 8 AM to sunset.

The campground has 85 sites, located among the trees on both sides of the creek. Also available are a group campsite, hike or bike campsites, and enroute camping. Ranger-led programs are offered. Maximum length for campers and RVs is 30 feet; for trailers, 27 feet. No hookups. Pay showers available. Dogs must be leashed at all times and kept in a tent or vehicle at night. Fee for day use and camping. Camping reservations for specific sites are available during all but the winter months; call: 1-800-444-7275. The campground is open year-round, except when winter rains cause flooding; sites are available first-come, first-served when reservations are not offered. For park information, call: 805-968-1033.

BIKE PATH: *Seaward of Hwy. 101, between Refugio and El Capitán State Beaches.* A paved path, fenced on the inland side, leads two and a half miles from Refugio State Beach to El Capitán State Beach. The path is located seaward of Hwy. 101 and the railroad tracks. Much of the bike path runs along the blufftop, but there are beach access points at intervals along its route. Beaches here are sandy, narrow, and backed by sandstone cliffs; scattered headlands form small coves.

EL CAPITÁN STATE BEACH: *12 mi. W. of Goleta.* Use exit 117 off Hwy. 101, and turn seaward for the park, which includes sandy and rocky beaches and upland day-use and camping facilities on the narrow coastal terrace. El Capitán Creek flows into the sea at a rocky point, near the park's entrance station. Day-use parking is located near the point, and a wheelchair-accessible, paved path leads along the shore to picnic tables sited in a lawn. The rocks at the point harbor tidepool organisms, and El Capitán Creek is bordered by a dense riparian woodland. Monarch butterflies congregate in the trees here. El Capitán Point offers decent surfing conditions occasionally, when winds and tides cooperate. The El Capitán Beach Store, offering snacks and beach equipment rentals, is open seasonally, next to the day-use parking lot. A beach wheelchair is available; call ahead to reserve.

To the west of the day-use parking lot is a long sandy beach. This is a popular swimming beach, and lifeguards are on duty during the summer months. The west end of the beach is backed by angled planes of sandstone, part of the Monterey Formation. Look for long-billed curlews feeding on the beach along the water's edge. Day use hours are from 8 AM to sunset.

The campground is located on the bluff above the beach. There are 142 family campsites and five group campsites scattered among oak trees; hike and bike sites and enroute camping are also available. Stairs provide access from the campsites to the sandy shore. Maximum length for trailers and RVs is 42 feet; no hookups are available, but there is a dump station. No fires except in fire ring. Dogs must be leashed at all times, kept in a tent or vehicle at night, and are not allowed on the beach. Fee for day use and camping; pay showers available. Camping reservations are recommended, and specific sites can be reserved; call: 1-800-444-7275. For park information, call: 805-968-1033.

Some 2,500 acres, formerly part of the old El Capitán Ranch, are slated to become part of the El Capitán State Beach through efforts spearheaded by the Trust for Public Land. Trails into the new holdings, which are inland of Hwy. 101, are available, but other facilities are yet to be developed.

EL CAPITAN RANCH CAMPGROUND: *12 mi. W. of Goleta.* El Capitan Ranch Campground shares exit 117 off Hwy. 101 with El Capitán State Beach; turn inland to reach the privately operated campground in a wooded canyon. There are 108 cabins, 26 furnished

El Capitán State Beach

El Capitán State Beach

tents, and six meeting spaces; day use also available. Facilities include a store, laundry, snack bar, and game arcade. Fifteen miles of hiking trails are accessible from the campground, including some within neighboring El Capitán State Beach. Activities include docent-led nature hikes on Saturday mornings, storytelling, and stargazing. Fees apply. For information and reservations, call: 805-968-2214.

NAPLES REEF: *1 mi. offshore, S.E. of Naples Point.* Accessible only by boat, the reef supports large numbers of surfperch, rockfish, lingcod, and cabezon, as well as kelp bass, California sheephead, Pacific barracuda, spiny lobster, and abalone. Dense giant kelp forests cover the reef, and the kelp understory provides shelter from predators for young fish. Water depths are between 30 and 60 feet. Naples Reef is an important sport and commercial fishing site. Boats from Santa Barbara harbor visit the area regularly; scuba divers collect lobster and abalone, and commercial divers harvest abundant sea urchins.

The cove at Refugio was for a while the chief contraband port on the southern California coast. José María Ortega, whose father had accompanied Portolá's 1769 overland expedition of discovery, established a compound here of adobe structures and raised wheat, grapes, and cattle. Ortega also engaged in trading hides, tallow, and wine with Yankee ships, at a time when Spain forbade commerce with foreigners. In 1818, word reached California from the Hawaiian Islands that a privateer by the name of Hippolyte de Bouchard was bent on mischief in Spanish California. Bouchard had been authorized by the fledgling nation of Argentina to seize Spanish ships and disrupt commerce in Spain's colonies. With two gunships and 250 men, Bouchard attacked and burned the settlement at Monterey and then appeared at the Ortega ranch at Refugio, where he laid waste to the adobe casa.

Joseph John Chapman, an American who had been pressed into service by Bouchard, escaped the gang; he later married Guadalupe Ortega and settled in Santa Ynez. The Refugio Road leads from the beach up the canyon and across the Santa Ynez Range; this was the route taken by 19th century traffic between Mission Santa Barbara and Mission Santa Inés, located in the town of Solvang.

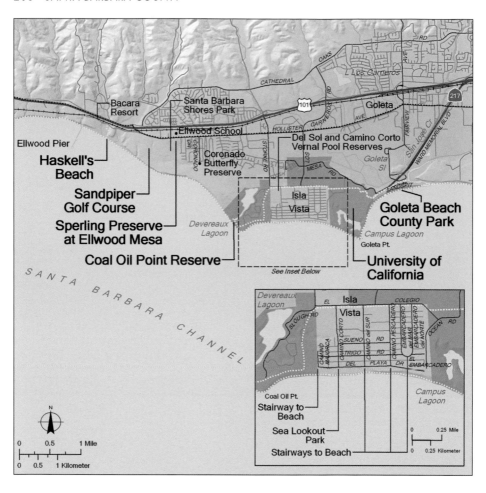

Haskell's Beach

Goleta Area

	Sandy Beach	Rocky Shore	Trail	Visitor Center	Campground	Wildlife Viewing	Fishing or Boating	Facilities for Disabled	Food and Drink	Restrooms	Parking	Fee
Haskell's Beach	•	•				•	•	•		•	•	
Sandpiper Golf Course									•	•	•	•
Sperling Preserve at Ellwood Mesa			•			•					•	
Coal Oil Point Reserve	•	•	•			•					•	
Stairway to Beach	•											
Sea Lookout Park	•											
Stairways to Beach	•											
University of California	•	•	•	•		•		•	•	•	•	•
Goleta Beach County Park	•					•	•	•	•	•	•	•

HASKELL'S BEACH: *W. end of Hollister Ave., Goleta.* A public beach access path and parking lot are at the east end of the Bacara Resort grounds, seven-tenths of a mile from Hollister Ave. A wheelchair-accessible, decomposed-granite path leads to the beach, restrooms with running water, and outdoor showers. The remnants of an old oil and gas pier were once located southeast of this beach; two wells produced oil and gas until 1958. In 2005, the Arco Company removed most of the pier and abandoned in place eight concrete caissons, which became the foundation for an artificial reef. The reef's purpose is to encourage the growth of marine life, including kelp, invertebrates, and game fish, such as calico bass, sand bass, rockfish, halibut, and perch. There are also four above-water platforms for cormorants and brown pelicans, which became accustomed to roosting on the old platform pilings.

Platform Holly, the only remaining oil production facility off Goleta in state-owned waters, is located almost due south. Other oil platforms off Santa Barbara County are located farther offshore under the primary jurisdiction of the federal government. Oil and gas seep naturally from subsurface deposits under the Santa Barbara Channel; an estimated 100 barrels surface daily less than a mile off of Coal Oil Point. Some of the natural seepage, mostly gas, is captured by a containment structure operated by Venoco, Inc.

Haskell's Beach

Coal Oil Point Reserve

SANDPIPER GOLF COURSE: *7295 Hollister Ave., Goleta*. This 18-hole course shares an access road with the Bacara Resort hotel. Some of the holes play along the ocean bluff, with spectacular views of the Santa Barbara Channel and the Channel Islands. The facility includes a pro shop and snack stand. Open to the public; entrance fee. For information, call: 805-968-1541.

SPERLING PRESERVE AT ELLWOOD MESA: *Hollister Ave., 1.5 mi. W. of Storke Rd., Goleta*. Opposite the Ellwood School is a parking lot providing access to a large open space area extending seaward to the ocean bluff. Equestrian, bicycle, and foot trails lead from the parking area into the Santa Barbara Shores Park and the neighboring Sperling Preserve, both managed by the city of Goleta. Brush rabbits, sparrows, and California towhees inhabit the undeveloped mesa, along with an occasional peregrine falcon and nesting raptors such as white-tailed kites, red-shouldered hawks, and Cooper's hawks. The Sperling Preserve is home to one of California's largest overwintering monarch butterfly aggregations, divided among several

groves. From the blufftop, there are views of the Santa Barbara Channel and steep, mostly unimproved paths down to the beach. For information, call: 805-961-7500. Public acquisition of open space property on Ellwood Mesa was supported in part by funds derived from offshore oil and gas operations in Santa Barbara County's coastal waters.

Near the end of Coronado Dr. is another entry point into the eucalyptus groves that harbor overwintering monarch butterflies. The Coronado Butterfly Preserve, managed by the Land Trust for Santa Barbara County, includes coastal sage scrub habitat and meadows; look for white-crowned sparrows and buckeye butterflies. Call: 805-966-4520. From the preserve, a trail leads into the monarch butterfly grove known as the Ellwood Main, with tens of thousands of butterflies during most winters. Beyond the grove, trails lead to the coastal bluffs and the beach.

COAL OIL POINT RESERVE: *Off Slough Rd., Isla Vista*. This reserve protects coastal dunes, a lagoon, and a beach where western snowy plovers nest, all increasingly scarce environments in southern California. The reserve is used primarily for educational and research purposes, but public access is available to parts of it. A mile-and-a-half-long trail circling the lagoon is open to pedestrians only; no dogs or bicycles. Park outside the reserve on El Colegio Rd. near Storke Rd., and walk in along Slough Rd. Interpretive panels are posted along Devereux Slough, which attracts thousands of migratory birds. Visitors in vehicles may also stop in small pullouts along Slough Rd., but drivers must stay with their vehicles. At Coal Oil Point, walkers may continue northwest along the beach to make a circuit of the Coal Oil Point Reserve's perimeter. The Pond Trail, which starts at the beach on the western side of Coal Oil Point, bisects the reserve and is also open to the public. Snowy plovers use the western side of Coal Oil Point, nesting there in the spring and summer months; beach visitors are limited to the wet sand area during this period.

STAIRWAY TO BEACH: *Camino Majorca at Del Playa Dr., Isla Vista.* Benches, a picnic table, and stairs to the West Campus Beach are located near Camino Majorca and Del Playa Dr. Look for sanderlings and sandpipers on the sandy beach, which extends one half mile west to rocky Coal Oil Point. Parking on the shoulder of Camino Majorca.

SEA LOOKOUT PARK: *Camino Corto at Del Playa Dr., Isla Vista.* A grassy blufftop park overlooks Isla Vista Beach. Facilities include a sand volleyball court, unusual wooden sunning and lounging decks, and picnic tables. The often-narrow beach is accessible via the Escondido Access stairway, located in mid-block, east of Sea Lookout Park. On-street parking only. Owned by Santa Barbara County and maintained by Isla Vista Recreation and Park District; call: 805-968-2017. To the west of the park, in an open space area on the bluff, is a small area of vernal pools, which once covered much of the flat oceanfront mesa on which Isla Vista is located. The Del Sol and Camino Corto Vernal Pool Reserves, much larger community parks featuring vernal pools restored by the Isla Vista Recreation and Park District, are located six blocks north on both sides of Camino Corto, near El Colegio Rd.

STAIRWAYS TO BEACH: *Camino del Sur, Camino Pescadero, and El Embarcadero Rd. at Del Playa Dr., Isla Vista.* Three stairways across from street ends lead to an often-narrow beach; on the bluff are Window to the Sea Park, Pescadero Park, and pleasantly landscaped Pelican Park. On-street parking only. Maintained by the Isla Vista Recreation and Park District; call: 805-968-2017. The narrow beach here is at least in part a result

Camino Majorca at Del Playa Dr., Isla Vista

of erosion from severe storms during the El Niño winter of 1982–1983. Much of the beach was lost at that time, and it has only partially recovered since then. The narrow beach has allowed ocean waves to quickly erode the bluff, undermining apartment units that were built too close to the bluff edge. Despite attempts at shoring up both the buildings and the eroding bluffs, buildings continue to be imperiled.

UNIVERSITY OF CALIFORNIA: *Off Clarence Ward Memorial Blvd., Goleta, and El Colegio Rd., Isla Vista.* Among California's institutions of higher learning, the University of California at Santa Barbara occupies a setting more marine than perhaps any other. The campus includes beaches, a coastal lagoon, and a variety of natural areas. The main campus at Goleta Point and the west campus at Coal Oil Point are separated by the residential community of Isla Vista, bordered by Isla Vista Beach. One-hour walking tours of the campus are offered weekdays at 12 noon and 2 pm; call: 805-893-8175. The University's Marine Science Institute offers aquarium and touch tank tours led by undergraduate docents. The Research Experience and Education Facility is open Tuesday through Thursday to school groups and on Saturday to the general public. No entrance fee; call: 805-893-8765.

Campus visitors may park in unmetered lots with a "C" designation; tune to AM radio frequency 1610 to hear a recording about parking on campus. A metered parking area for public coastal access is located at the end of Ocean Rd. off El Colegio Rd. A wheelchair-accessible path leads along the blufftop past picnic tables and through an area of restored coastal prairie and vernal pools with interpretive panels. Adjacent to the Campus Lagoon is the University Center, with dining facilities, the UCSB Bookstore, and a visitor information counter; call: 805-893-2487. To

Campus Lagoon, University of California at Santa Barbara

reach Parking Services, call: 805-893-2346. The campus is served by Metropolitan Transit District buses; call: 805-683-3702

GOLETA BEACH COUNTY PARK: *5990 Sandspit Rd., Goleta.* Popular Goleta Beach County Park, located at the mouth of Goleta Slough, offers a range of recreational facilities, including a fishing pier, children's play area, outdoor shower, volleyball courts, platforms overlooking the slough for observing wildlife, and a sandy beach sheltered by Goleta Point. In 2003-04, about 80,000 cubic yards of sand were added to the beach to counter the continuing erosion of the shoreline, and the county of Santa Barbara is exploring long-term options for protecting both the grassy park and the sandy beach. Lifeguards are on duty during the summer months. A beach wheelchair is available; check with a lifeguard or other park personnel. A restaurant and snack stand are located next to the pier. Group picnic areas can be reserved; for information, call: 805-568-2465.

War seems far away from the Goleta coast, a haven of sandy beaches and bluffs frequented by trail walkers, surfers, golfers, and college students taking a break from their books. Not so during World War II, when a Marine Corps Air Station occupied much of Goleta—including the Santa Barbara Municipal Airport and the University of California at Santa Barbara campus—and the Ellwood oil fields were the site of an Imperial Japanese naval attack.

Just past sunset on February 23, 1942, an I-17 submarine surfaced offshore of what is now the Sandpiper Golf Course and volleyed 16 shells at the Ellwood oil fields. The oil fields, which extended from Haskell's Beach to Coal Oil Point, were one of the most important oil deposits on the coast and contained numerous wells, piers, and other facilities, both on and offshore.

Three shells hit near the Bankline Oil refinery and damaged a pier and an oil well derrick. Another shell exploded on Tecolote Ranch, three miles inland, but the majority fell short of land and sank in the channel. The bombing lasted about 20 minutes and caused approximately $500 in damage. The attack was the first foreign bombing of the American mainland since 1812 and one of only five attacks on the mainland to occur during the war. Although little physical damage was incurred, the bombing shattered many Americans' sense of security and heightened the fear of attack on home soil. A marker commemorating the attack is found past the ninth green on the Sandpiper Golf Course.

Vernal Pools

VERNAL POOLS occur in shallow depressions underlain by an impermeable hardpan that prevents water from draining following winter rains. The combination of standing water for long periods during the winter and spring and total desiccation during the hot, dry Mediterranean summer restricts the vernal pool biota to a special group of species adapted for just such extreme conditions and excludes many of their potential upland and wetland competitors. The pools support specialized animal life, such as tiny shrimp that survive the summer drought in their egg stage, and many small, mostly annual plants, which flower as the water in the pools begins to evaporate. In the spring, this pattern of blooming often results in concentric rings of showy flowers as species adapted to different amounts of moisture flower sequentially. Green and blooming in spring, vernal pools may have a lifeless appearance by the end of a dry summer, but they come to life again as annual rains fall. Vernal pools are rare habitats found in California and few other places.

Isla Vista is built on a flat coastal plain that once was dotted with vernal pools. Many were built upon, as the residential community grew, or were altered as part of mosquito abatement activities. A few vernal pools remain, and some have been restored. Isla Vista may be the only town in the world to have a vernal pool community park celebrating this aspect of its natural history.

Coyote thistle

Coyote thistle (*Eryngium vaseyi*) is the only green, prickly plant that occurs in vernal pools. It is a member of the *Apiaceae*, or carrot, family, and like most members of this family it grows a long, fleshy taproot. The attractive, tiny purple flowers open in the late spring. This interesting plant has two different forms: when it is growing in water, the leaves are hollow and deliver air to the roots; when growing in a dry location, the leaves are solid and bristly like a thistle. Look for both life forms when you visit the vernal pools.

Downingia

Downingia (*Downingia concolor*) is one of the best-known vernal pool plants. Its diminutive purple flowers provide a beautiful splash of color along the edges of vernal pools during the spring. Downingia are annual plants; an annual is a plant that germinates, flowers, and dies in the course of one year. These well-adapted vernal pool endemics germinate while the seeds are under water. As the pools dry, downingia provide a stunning purple display. Downingia are pollinated by native solitary bees. Solitary bees dig burrows where they nest in the upland areas adjacent to vernal pools.

Western spadefoot toads (*Spea hammondii*) gather at vernal pools to breed in early spring, after winter rains fill the pools. The females lay several hundred jelly-covered eggs in small clusters on plants, just below the surface of the water. Spadefoot toads escape the summer's heat by burying themselves in the moist soil. By early summer, the pools are dry and this species goes into aestivation, or summer hibernation. Spadefoot toads have a small, black "spade," or shovel-like growth, on the first toe of each hind foot. This hardened tissue allows them to dig in loose soil to create their burrows. Their genus name *Spea* means "cave," which refers to this toad's habit of spending its adult life in burrows. If picked up, the western spadefoot toad may secrete a substance that can inflame the skin or cause hayfever symptoms like runny nose and watery eyes.

Western spadefoot toad

Water pygmyweed (*Crassula aquatica*) is a tiny, annual, aquatic plant commonly found in vernal pools. It is a type of succulent with short, fleshy leaves. The genus name, *Crassula*, comes from the Latin word *crassus* meaning "thick, dense, or fat," in reference to the succulent leaves of this genus. The species name *aquatica* comes from the Latin word *aquaticus* meaning "living or found in or by the water." The leaves of water pygmyweed are only about one-quarter inch long. You might have to get on your hands and knees to find this cute little plant in the shallow waters and bare mud. Look for the white, four-petaled flowers in the leaf axils, or axis where the leaf and stem meet.

Water pygmyweed

Camino Corto Vernal Pool Reserve, Isla Vista

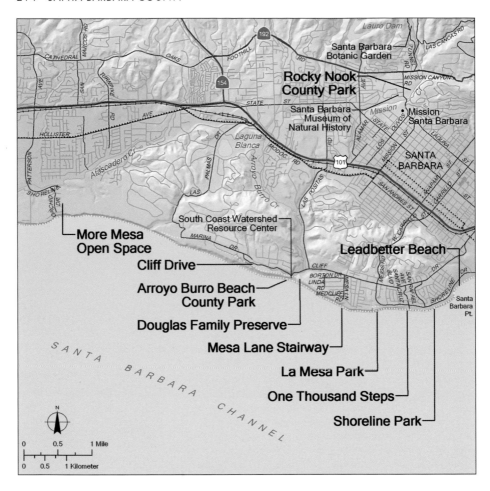

Arroyo Burro Beach County Park

Santa Barbara West

	Sandy Beach	Rocky Shore	Trail	Visitor Center	Campground	Wildlife Viewing	Fishing or Boating	Facilities for Disabled	Food and Drink	Restrooms	Parking	Fee
More Mesa Open Space			•			•						
Cliff Drive												
Arroyo Burro Beach County Park	•						•		•	•	•	
Douglas Family Preserve			•			•						
Mesa Lane Stairway	•											
La Mesa Park										•	•	
One Thousand Steps	•	•										
Shoreline Park	•					•	•			•	•	
Leadbetter Beach	•						•	•	•	•	•	
Rocky Nook County Park			•					•		•	•	

MORE MESA OPEN SPACE: *S. Patterson Ave., off Orchid Dr., Santa Barbara.* This undeveloped open area includes oak woodland and riparian habitat. There are trails, but no other facilities. Native plant restoration has taken place along Atascadero Creek. Maintained by Santa Barbara County; for information, call: 805-568-2460.

CLIFF DRIVE: *W. of Arroyo Burro Beach to Marina Dr., Santa Barbara.* Scenic drive with views of the Santa Barbara Channel and Channel Islands; there is a turnout near Yankee Farm Rd.

ARROYO BURRO BEACH COUNTY PARK: *Off Cliff Dr., .2 mi. W. of Las Positas Rd., Santa Barbara.* This popular park, known locally as Hendry's Beach, is located at the mouth of a narrow coastal canyon. A beach of fine sand extends both west and east at the mouth of Arroyo Burro Creek. Lifeguards on duty during the summer. A popular restaurant and snack stand overlooks the sand; a small grassy picnic area for families and groups is nearby, with tables and barbecue grills. Outdoor showers. Open 8 AM to sunset; for reservations or information, call: 805-568-2460. The two parking lots are often crowded on summer days. Metropolitan Transit District bus line #5 stops at the park entrance; for information, call: 805-683-3702. Arroyo Burro Beach County Park is the site of the South Coast Watershed Resource Center, which offers educational programs about the connection between healthy watersheds and everyday activities such as proper disposal of household chemicals.

Arroyo Burro Creek provides estuarine habitat for the endangered tidewater goby, and shorebirds feed near the mouth of the creek. Restoration measures for the estuary, funded by the State Coastal Conservancy and other entities, are planned for late 2006, enhancing the habitat for fish, birds, and riparian plants and improving water quality. The ripe fruit of the lemonade berry shrub, which grows along the bank of the creek, was used by Indians and early settlers to make a cooling drink. Excavation of a well-preserved Chumash midden east of the park produced shellfish remains, bones, and artifacts such as flake tools, hammerstones, and projectile points.

DOUGLAS FAMILY PRESERVE: *2551 Medcliff Rd., Santa Barbara.* A 70-acre open space area borders the east side of Arroyo Burro Beach County Park. Once known as the Wilcox property and planned for residential development, the property was donated to the City of Santa Barbara by community members, including the family of actors Kirk and Michael Douglas. From the 150-foot-high flat mesa, which rises sharply above surrounding land, there are commanding views of the Santa Barbara Channel and the Channel Islands, as well as the Santa Ynez Mountains to the north. Oak woodland covers the northern slopes of the preserve, providing roosting and nesting habitat for raptors, and coastal bluff scrub vegetation grows on the slopes facing the beach. There is a trail entrance, but no parking, on Cliff Dr. opposite Las Positas Rd. Other entrances with on-street parking are located at the ends of Borton Dr., Linda Rd., and Medcliff Rd. No restrooms or facilities. Open sunrise to 10 PM; dogs permitted off-leash.

MESA LANE STAIRWAY: *S. end of Mesa Ln., Santa Barbara.* A long stairway descends a steep cliff to a beach popular with sunbathers and surfers; limited on-street parking. This is the only access to the beach for a mile in either direction. Maintained by the City of Santa Barbara; open sunrise to 10 PM. For information, call: 805-564-5437.

LA MESA PARK: *Meigs Rd. at Shoreline Dr., Santa Barbara.* This pleasant, grassy neighborhood park overlooks a wooded ravine and a glimpse of blue water; there is no beach access. Facilities include a children's playground and a group picnic area that can be reserved. Open sunrise to 10 PM. Call: 805-564-5437. In 1856 a 30-foot lighthouse was built on top of the bluff, near the park. Destroyed by the 1925 earthquake that devastated Santa Barbara, the lighthouse was replaced by an automated light, which is not accessible to the public.

ONE THOUSAND STEPS: *End of Santa Cruz Blvd., Santa Barbara.* Beach access stairway, originally constructed in 1923. Very limited street parking, on the north side of Shoreline Dr.; no facilities. The beach is narrow; at low tide, there are tidepools along the shore.

Leadbetter Beach

SHORELINE PARK: *Shoreline Dr. between San Rafael Ave. and La Marina Dr., Santa Barbara.* This 15-acre landscaped neighborhood park offers a wide range of activities, including whale watching from the high bluff, kite-flying, Frisbee, and picnics. Facilities include a large lawn, children's playground, wheel-chair-accessible walking paths, restrooms, and a sister-city Japanese garden. Two off-street parking lots; a stairway to the narrow sandy beach is tucked among trees, in the middle of the linear park. At the eastern end of the park, there are panoramic views of Santa Barbara Harbor, the city of Santa Barbara, and the Santa Ynez Mountains beyond. Call: 805-564-5437.

LEADBETTER BEACH: *Shoreline Dr. at Loma Alta Dr., Santa Barbara.* A wide stretch of sandy beach is located along a shallow cove, between Santa Barbara Point and the Santa Barbara Harbor breakwater. A popular wind-surfing spot, Leadbetter Beach was created in 1929 after construction of the harbor breakwater caused large amounts of sand to be deposited upcoast from the breakwater. The parking lot is shared with Santa Barbara Harbor to the east; a snack stand, restaurant, and landscaped picnic area with barbe-cue grills are adjacent to the parking area. A beach volleyball court can be reserved. At low tide, pedestrian access to Leadbet-ter Beach is also possible along the sand at the base of the bluff from the Shoreline Park

stairway; there are tidepools near Santa Barbara Point. Call: 805-564-5437.

ROCKY NOOK COUNTY PARK: *610 Mission Canyon Rd., Santa Barbara.* Bordering Mission Creek is an oak-shaded park featuring family and reservable group picnic areas, barbecue grills, short trails, and playground equipment. Open 8 AM to sunset; for information, call: 805-568-2460.

Marbled godwit at Leadbetter Beach

Inland of coastal Hwy. 101 are visitor attractions such as the Santa Barbara Botanic Garden in Mission Canyon; Skofield Park at 1819 Las Canoas Rd., Santa Barbara, with trail links to the Los Padres National Forest; and Toro Canyon County Park, on Toro Canyon Rd. in Montecito. Passing near these and other inland parks is Route 192, an east-west scenic route that follows the base of the Santa Ynez Mountains. Beginning at the San Marcos Pass highway, Route 192 twists and turns eastward among the foothills, offering views of the mountains and occasional glimpses of the sea. On the slopes above Summerland and Carpinteria, the route passes through avocado and citrus groves, before reaching the Ventura County line.

Route 192 takes numerous names along its route, including Foothill Rd., Sycamore Canyon Rd., and East Valley Rd., but the numbered highway markers are prominent all the way.

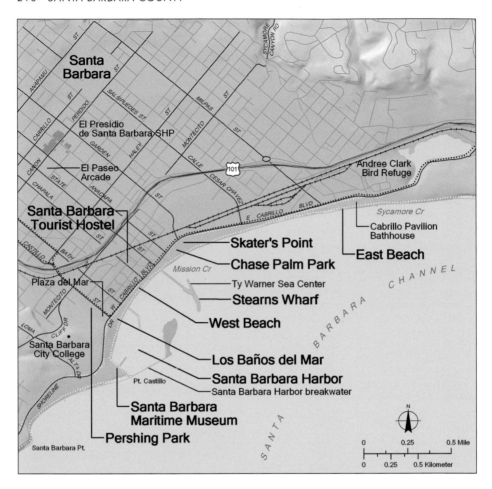

Stearns Wharf

Santa Barbara Central

	Sandy Beach	Rocky Shore	Trail	Visitor Center	Campground	Wildlife Viewing	Fishing or Boating	Facilities for Disabled	Food and Drink	Restrooms	Parking	Fee
Santa Barbara	•	•	•	•		•	•	•	•	•	•	
Pershing Park									•	•		
Santa Barbara Maritime Museum				•					•			•
Santa Barbara Harbor				•		•	•	•	•	•	•	•
Los Baños del Mar									•	•	•	
Santa Barbara Tourist Hostel									•	•		
West Beach	•		•			•	•		•	•		
Stearns Wharf				•		•	•	•	•	•	•	•
Chase Palm Park	•		•					•	•		•	
Skater's Point								•				
East Beach	•		•			•	•	•	•	•	•	

SANTA BARBARA: *90 mi. N.W. of Los Angeles.* Santa Barbara became a visitor destination early on; the Victorian-style Upham Hotel, opened in 1871, is perhaps the oldest continuously operating hotel in Southern California. The Arlington Hotel, opened in 1875 on State St. where the Arlington Theater now stands, helped establish Santa Barbara as a prime winter resort. President Theodore Roosevelt; David Kalakaua, the last king of Hawaii; and Princess Louise, daughter of Britain's Queen Victoria, stayed at the Arlington. Grandest hotel of all, perhaps, was the 600-room Potter, which stood on West Cabrillo Blvd., with the beach before it and the train station behind it, from 1902 until it burned in 1921.

The visitor center on E. Cabrillo Blvd. opposite Stearns Wharf offers a wealth of information. A highlight of a visit to Santa Barbara is the Santa Barbara Museum of Natural History, located at 2559 Puesta del Sol Rd. The museum features exhibits on the geology, zoology, and ethnographic history of the southern and central California coast and the Channel Islands. A complete blue whale skeleton is on display; tree-shaded gardens surround Spanish-style buildings. The museum provides a marine educational program, known as Los Marineros, to all fifth graders in Santa Barbara.

PERSHING PARK: *100 Castillo St., Santa Barbara.* This city park, a block from the beach, has softball and baseball fields and eight tennis courts. Home of the Santa Barbara City College tennis team, the courts are open to the public on weekends and after 5 PM on weekdays. Four tennis courts are lighted for play until 9 PM Monday through Friday. Restrooms are not wheelchair accessible. At the north end of the park is the Old Spanish Days Carriage Museum, featuring historic

Blue whale skeleton, Santa Barbara Museum of Natural History

Santa Barbara offers an electric shuttle service along State St. and the waterfront. The Santa Barbara Metropolitan Transit District's Line 22 serves the Santa Barbara Mission, County Courthouse, Museum of Art, Museum of Natural History, and, on weekends, the Botanic Garden in Mission Canyon. Other lines serve beaches and parks, some of which have very limited parking. Visit the Transit Center at 1020 Chapala St., Santa Barbara, or call: 805-683-3702.

horse-drawn vehicles; open weekdays, 9 AM to 3 PM, and Sundays, 1 PM to 4 PM. For information, call: 805-962-2353. Between Pershing Park and Cabrillo Blvd. is Plaza del Mar, a shady green oasis with a band shell.

SANTA BARBARA MARITIME MUSEUM: *113 Harbor Way, Santa Barbara Harbor.* The museum offers a wide range of exhibits about diving, whaling, yachting, surfing, and other sea-related topics in a historic building in the heart of Santa Barbara Harbor. The new Survival at Sea exhibit includes interactive features that allow visitors to experience driving a boat in the harbor and charting a course across the Santa Barbara Channel.

Navigation tools used in the past are on display. The Santa Barbara Maritime Museum offers educational programs for young people and public lectures on marine topics.

The Outdoors Santa Barbara Visitor Center is located on the fourth floor of the building housing the Maritime Museum. A small outdoor deck offers a magnificent view of the harbor scene; exhibits and information are available about the Los Padres National Forest, Channel Islands National Marine Sanctuary, Channel Islands National Park, and the City of Santa Barbara.

SANTA BARBARA HARBOR: *W. Cabrillo Blvd. at Shoreline Dr., Santa Barbara.* Harbor facilities include a marina with 1,000 boat slips, marine specialty shops, boat repair services including a dry dock, boat rentals, boat hoists with up to 3,000-pound capacity, fuel dock, and launching ramp. Fees apply for use of the hoist and launching ramp. Guest slips are available through the harbormaster; call: 805-963-1737. The harbor also offers restaurants with views of the boats, shoreline, and mountains; fishing supplies; and fishing and wildlife-viewing excursions. Scenic views are available from the walkway along the breakwater; good surfing conditions outside the harbor. Fees required for some spaces in the parking lot.

LOS BAÑOS DEL MAR: *400 block of W. Cabrillo Blvd., Santa Barbara.* Located next to West Beach, Los Baños del Mar offers an outdoor

Ty Warner Sea Center, Stearns Wharf

Santa Barbara Harbor

50-meter-long swimming pool, with daily lap and open swimming. Lifeguards on duty; lockers available. Parking shared with Santa Barbara Harbor. For hours and fees, call: 805-564-5418. Adjacent to the swimming pool is the West Beach Wading Pool, an 18-inch-deep pool for children up to seven years of age. The wading pool is open daily, weather permitting, from noon to 5 PM, May through September. Lifeguard on duty.

SANTA BARBARA TOURIST HOSTEL: *134 Chapala St., Santa Barbara.* Located across from the Amtrak station; 60 beds, laundry facilities, kitchen, showers, lockers, and free breakfast and Internet access. Bicycles and bodyboards available for rent. Open 8 to 11:30 AM, and 3:30 PM to midnight. For information, call: 805-963-0154.

WEST BEACH: *Seaward of W. Cabrillo Blvd., between Santa Barbara Harbor and Stearns Wharf.* A third of a mile of fine sand for sunbathing and beach volleyball; kayakers and wind surfers use this beach, which has relatively little wave action, to gain access to the waters of the harbor. A paved path for walkers, skaters, and bicyclists borders the beach seaward of palm-lined Cabrillo Blvd.

STEARNS WHARF: *Foot of State St., Santa Barbara.* When first completed in 1872 by local lumber merchant J. P. Stearns, the wharf was the longest deepwater pier on the California coast between San Francisco and Los Angeles. Stretching 1,500 feet long in 40 feet of water, the pier served both passengers and freight. After the pier was built, the practice of "lightering" was ended. This was a marine taxi service that took passengers, for about 50 cents apiece, from ships anchored beyond the surf line to the shore. People were loaded into surfboats, or "lighters," and rowed through the surf where they would then be carried ashore on the backs of sailors. The wharf was severely damaged by a storm in 1878 and a fire in 1973 and rebuilt both times. The wharf is used now for fishing, both commercial and recreational, and sightseeing. Bait and tackle, parasailing, and whale watching are offered on the wharf, along with shops and restaurants.

The Ty Warner Sea Center offers opportunities to experience and learn about the ocean in an over-water setting on Stearns Wharf. An expansion project completed in 2005 added an interactive shark exhibit, and a "live dive" program is planned, allowing visitors to speak with divers as they explore the sea floor. The Ty Warner Sea Center is part of the Santa Barbara Museum of Natural History; call: 805-962-2526.

Five hundred feet east of Stearns Wharf is the small estuary formed by Mission Creek, which meanders across the sand before flowing into the harbor. The creek originates on the south slopes of the Santa Ynez Mountains and winds along a six-mile course through the center of downtown Santa Barbara. Sections of the stream and its tributaries support remnant stands of riparian vegetation, including western sycamore and coast live oak trees. The Rattlesnake Trail leads from Skofield Park on Las Canoas Rd. into the creek's upper watershed. Although an urban creek, Mission Creek offers opportunities for restoration of habitat and scenic values; the City of Santa Barbara has improved a portion of the creek in an award-winning project, funded by the California Resources Agency and the Department of Parks and Recreation, at Bohnett Park, at West Anapamu and San Pascual Streets.

CHASE PALM PARK: *Both sides of E. Cabrillo Blvd., E. of Stearns Wharf, Santa Barbara.* The epitome of the Southern California shoreline park: improbably tall palm trees, sailboats bobbing on an azure sea, and the chaparral-covered Santa Ynez Mountains rising over all. A paved bicycle path extends the length of the park on the ocean side of Cabrillo Blvd. A perennially popular arts and crafts show is set up on Sundays and holidays among the palm groves. A multi-purpose grass area is the scene of drop-in games of soccer, as well as weekend tournaments. On the inland side of Cabrillo Blvd. are a nautical-themed children's playground, carousel, and snack stand. A wetland, winding paths, and fountains are also located here. Free concerts are held on the grass in summer. Call: 805-564-5418.

Chase Palm Park

East Beach

SKATER'S POINT: *E. Cabrillo Blvd. at Garden St.* A 14,600-square-foot cement skate park is located opposite the foot of Garden St. The skate park offers a half-pipe, rails, fun boxes, and other skating features. Open from 8 AM until one-half hour after sunset. For information, call: 805-564-5418.

EAST BEACH: *Seaward of Chase Palm Park, E. of Stearns Wharf, Santa Barbara.* This mile-long, city-maintained beach features 18 sand volleyball courts near its east end. The historic Cabrillo Pavilion Bathhouse located nearby at 1119 East Cabrillo Blvd. offers showers, lockers, and a weight room that are open to the public; fees apply. A beach wheelchair is available; volleyballs for rent. Call: 805-897-2680. The Cabrillo Pavilion building also houses an art center, a grill restaurant overlooking the beach, and restrooms.

Skater's Point

In Santa Barbara, many recreational outfitters are located around Stearns Wharf, at Leadbetter Beach, in Santa Barbara Harbor, and on lower State St. Dive shops and surf shops are on Anacapa St. near Cabrillo Blvd., and others are located elsewhere in Santa Barbara and in Isla Vista, Goleta, and on Linden Ave. and Santa Claus Ln. in Carpinteria. Bicycle shops are located all over, including Isla Vista, on Hollister Ave. in Goleta, and on Carpinteria Ave. in Carpinteria. Visitor maps and brochures provide listings of recreational outfitters.

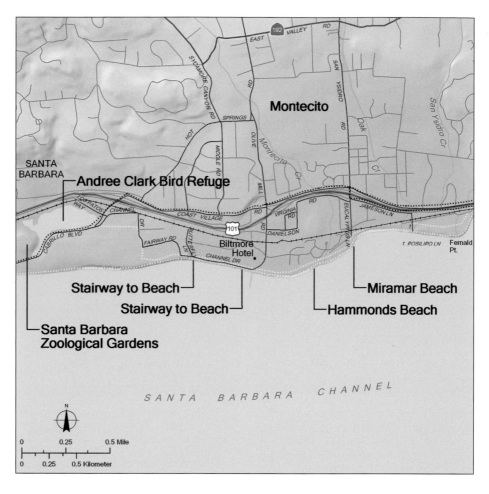

Andree Clark Bird Refuge

Santa Barbara to Montecito

	Sandy Beach	Rocky Shore	Trail	Visitor Center	Campground	Wildlife Viewing	Fishing or Boating	Facilities for Disabled	Food and Drink	Restrooms	Parking	Fee
Santa Barbara Zoological Gardens			•	•		•		•	•	•	•	•
Andree Clark Bird Refuge			•			•					•	
Stairway to Beach	•											
Stairway to Beach	•											
Hammonds Beach	•	•	•								•	
Miramar Beach	•										•	
Montecito	•	•	•				•		•	•	•	

SANTA BARBARA ZOOLOGICAL GARDENS:
500 Ninos Dr., Santa Barbara. A beautifully landscaped 81-acre zoo and park, located on a knoll overlooking East Beach and the harbor area. Facilities include a carousel, picnic area, snack stand, and a nature theater offering educational performances for children. In addition to snow leopards, western lowland gorillas, and other species from around the world, the zoo's collection includes the island fox, California sea lion, and barred tiger salamander, all native to the California coast. A California nature walk, next to the Andree Clark Bird Refuge, features native plants. Open daily, 10 AM to 5 PM, except for Thanksgiving, Christmas, and special events. Separate fee for entry and parking. Strollers, wagons, and wheelchairs are available for rent. For recorded information, call: 805-962-6310.

ANDREE CLARK BIRD REFUGE: *E. Cabrillo Blvd. at Hwy. 101, Santa Barbara.* A brackish marsh in a scenic setting provides habitat for a variety of birds, including herons, egrets, ducks, and geese. In summer, cormorants can be seen flying back and forth between the refuge's lake and the nearby harbor. There are several small islands in the lake. A grassy area along the south and east shores overlooks the refuge and incorporates a segment of the city's heavily used, shoreline bike path. Limited parking available on Los Patos Way. Interpretive signs are placed along a path along the north side of the refuge. Open from sunrise until 10 PM.

Richard Dana, who visited Santa Barbara in January 1835, wrote that the town and mission lay "on a low plain, but little above the level of the sea, covered with grass, though entirely without trees…" The surrounding hills were treeless too, Dana noted, as a result of "a great fire which had swept them off." Today, the arboreous appearance of Santa Barbara's streets and parks is one of its great attractions. The climate and soil support many California native and exotic trees, from temperate and subtropical climes. The famous Moreton Bay fig (*Ficus macrophylla*) at Chapala and Montecito Streets stands 76 feet high and has a crown width of 176 feet. The tree was reportedly planted as a seedling in 1877 by 9-year-old Adeline Crabb and her mother, Hannah. The book *Trees of Santa Barbara*, available at the Santa Barbara Botanic Garden, describes the trees of the region and notes where they can be seen. The Botanic Garden, located at 1212 Mission Canyon Rd., is a repository solely of California native plants.

STAIRWAY TO BEACH: *Butterfly Ln. at Channel Dr., Montecito.* A public stairway leads to the sandy beach, opposite the end of Butterfly Ln. On-street parking only; no facilities.

STAIRWAY TO BEACH: *Channel Dr., seaward of Biltmore Hotel, Montecito.* Channel Dr. is a scenic route with views of the Santa Barbara Channel. Approach Channel Dr. from Olive Mill Rd., because part of Channel Dr. is one-way, westbound. A public stairway leads to the narrow sandy beach. On-street parking; no facilities.

HAMMONDS BEACH: *End of Eucalyptus Ln., Montecito.* At the foot of Eucalyptus Ln. off San Ysidro Rd. is a small parking area; look for the trail between the high hedges of neighboring residences to reach Montecito Creek, a five-minute walk. A bridge over the creek provides access to trails down both sides of the waterway to the narrow beach, part of it rocky and part sandy.

MIRAMAR BEACH: *E. of Eucalyptus Ln., Montecito.* A narrow beach, accessible via ramp and trail at the end of Eucalyptus Ln., is located next to the old Miramar Hotel, which closed in 2000. Limited on-street parking. Access to the east end of the same beach is available at Posilipo Ln.; no nearby parking available, but roadside parking is allowed on Jameson Ln., west of Posilipo Ln.

MONTECITO: *2 mi. E. of Santa Barbara.* Montecito originated as a health resort that drew visitors from America's East Coast and beyond. The Montecito Hot Springs Hotel was built in the late 19th century in a canyon east of Santa Barbara for those who sought its mineral waters. Today, the community is largely residential, and many of Montecito's architectural and botanical treasures are hidden from public view behind hedges and walls. But one of the most dramatic, the estate known as Lotusland, is open for public tours. The 37-acre estate includes gardens devoted to colorful, dramatic plants, many of exotic origin. The gardens, some almost fantastic in appearance, contain extensive collections of palms, cactus, bromeliads, ferns, and cycads. There are also a Japanese garden and a butterfly garden. The gardens reflect the taste for the striking and the beautiful of Polish-born Madame Ganna Walska,

an actress and singer turned garden creator, who owned the estate from 1941 until her death in 1984. Docent-led tours are held between mid-February and mid-November. Fee for the tour; reservations required. For information, call: 805-969-9990 on weekdays, between 9 AM and noon.

Santa Barbara's appearance is distinctive for its Spanish-themed architecture. From the start, simple whitewashed-structures were built of sun-dried adobe blocks, the most practical building material in an area without close-by forests. After Stearns Wharf was built, wood frame construction became more popular, and Victorian-style buildings appeared. But the Spanish theme was continued by Mission revival-style buildings and Spanish Colonial and Mediterranean styles exhibited by the El Paseo Arcade, built in 1922–23 at 800 Anacapa St. The arcade takes the form of a Spanish village, with shops and offices grouped around pedestrian passageways, incorporating the original De La Guerra home built in 1819–1826. Santa Barbara took on an even more Spanish appearance following the great earthquake of June 25, 1925, which demolished much of the city's downtown area. A Board of Architectural Review promoted a consistent Spanish-themed architecture, and in 1960 the city created the El Pueblo Viejo District around the old Presidio area to enforce a consistent architectural theme. But walk along the side streets off lower State St. to see many wood-framed Victorian cottages and other 19th and early 20th century building styles, which survived the 1925 earthquake and look much like those in other 19th century California towns.

Oil is a sticky substance, hard to clean from beaches and harbors, hard to remove from feathers and fur. Images of oil spills—blackened beaches, oil-soaked birds—are hard to erase from memory. The most enduring aspect of the 1969 Santa Barbara oil spill, however, may be the environmental programs, agencies, and laws that it helped to create.

On January 28, 1969, a blowout occurred at Union Oil's Platform A, about six miles southeast of downtown Santa Barbara. The disaster resulted from a pressure imbalance in a well that had an inadequate protective casing. Highly pressurized natural gas and oil shot out of the well, the oil's geologic formation ruptured, and 4.2 million gallons of heavy crude oil poured out through five long faults in the ocean floor. The oil spread in an 800 square mile slick and fouled beaches from Pismo Beach to the Mexican border.

As the oil moved onshore, so did the bodies of dolphins, seals, sea birds, and other marine life; an estimated 3,686 sea birds are known to have perished. Gray whales avoided the area on their annual migrations, and shore birds stayed away. The spill shut down local tourist, boating, and commercial fishing industries, and damaged marinas, park facilities, and beachfront homes and resorts, resulting in millions of dollars in property damage and lost revenue. The tangible horror of the spill and its sweeping impacts united the local community, as civic leaders and business owners worked alongside student activists and homemakers on clean-up efforts and the establishment of environmental organizations and programs.

Widespread publicity brought the spill into the nation's living rooms, fueling support for the burgeoning environmental movement and thrusting environmental protection to the forefront of the national agenda. The spill helped inspire a new era of environmental activism and government regulation that resulted in the formation of the U.S. Environmental Protection Agency and adoption of important federal and state legislation, including the National Environmental Policy Act of 1970, the federal Coastal Zone Management Act, the California Environmental Quality Act, and the California Coastal Act of 1976.

Platform A can be seen on the horizon from Stearns Wharf in Santa Barbara.

Platform A

Santa Ynez Mountains

© 2007, Tom Killion

Geology of the Transverse Ranges

THE SANTA YNEZ, San Rafael, and Santa Monica Mountains, together with their more inland counterparts, the San Bernardino and San Gregorio Mountains, collectively constitute the Transverse Ranges physiographic province. The northern Channel Islands, really the tops of yet another east-west range that is mostly submerged, are also part of the Transverse Ranges province as are, somewhat paradoxically, the deep valleys separating these ranges and the Los Angeles Basin.

A glance at a map reveals the obvious reason for the name for the province—these roughly east-west trending mountains and valleys run at almost right angles to most of California's other mountain ranges, which generally trend from northwest to southeast. The geologic evolution of this province has been the subject of ongoing study and debate among geologists for decades.

The simplest part of the geologic history of the province to unravel is the relatively recent past—the last four million years or so. During that time, the prevailing theme in the evolution of the area has been one of compression of the earth's crust. The cause of this compression has to do with the relative plate motions between the North American and Pacific plates and the orientation of the San Andreas fault. The regional trend of the fault in central California is northwest-southeast, but a segment in the Tejon Pass area trends much closer to east-west. This "Big Bend" in the San Andreas fault results in a regional zone of compression. As the Pacific and North American plates slide past each other along the fault, they are forced to converge in this region, causing compression and uplift.

This compression and uplift is manifested in broad upwarped folds of the earth's crust, called anticlines, which form the cores of several of the area's mountain ranges—notably the coastal Santa Ynez and Santa Monica Mountains. Some of these anticlines are growing upward at amazing rates, causing the mountains to grow. The areas between the anticlines—where the rocks are folded downward—are called synclines, which form deep basins that are filled with sediment eroded from the rising mountains. Such basins underlie Los Angeles, the Ventura-Oxnard Plain, the Santa Barbara Channel, the Santa Ynez Valley, and the Santa Maria Valley. All of these mountain ranges and basins are cut by numerous faults. Most of the mountain ranges and basins have fault boundaries, and faults within the basins break them into a series of sub-basins, sometimes with very complex geologic histories. These young basins contain immense quantities of sediment—the Los Angeles Basin, for example, has been filled with 18,000 feet of sediment in the past 4 million years—an astonishingly high sedimentation rate to match the remarkable uplift rates of adjacent ranges.

The early history of the Transverse Ranges is much more poorly understood than the recent history. The sedimentary rocks that have been uplifted to form the mountain ranges were deposited mostly in marine environments beginning about 80 million years ago, although a period of uplift allowed for the deposition of alluvial fan sediments briefly around 20 to 30 million years ago, before marine conditions returned once again. When they did, the extensive Monterey Formation, consisting of shales and siltstones containing abundant organic matter and the siliceous remains of diatoms, was deposited. The Monterey Formation also contains numerous layers of altered volcanic ash, derived from a series of volcanoes fed by the Conejo volcanic series—igneous rocks that cut through slightly older sedimentary rocks in the Santa Monica Mountains. Most of the oil in coastal southern California originated in the Monterey Formation.

Santa Ynez Mountains

A number of geologic anomalies in the Transverse Ranges are best explained if one considers much of the Transverse Ranges province to be a block of crust that has been transported from the south and rotated into its current position. Evidence for the rotation comes from studies of paleomagnetism—records of the Earth's past magnetic field that can be extracted from certain rocks. The direction of magnetic north preserved in the rocks only makes sense if the block has been rotated by as much as 100 degrees; younger rocks record progressively smaller amounts of rotation, consistent with their deposition as rotation continued. Also, there are distinctive rock types—such as the Poway Conglomerate with its very distinctive red rhyolite pebbles—that are found in southern California and in parts of the Transverse Ranges, but nowhere else. Assuming that these rocks were once contiguous, northward transport and rotation of the Transverse Ranges block is necessary to explain their present distribution. Unlike the recent history of the Transverse Ranges, the period from about 19 to 5 million years ago was one of general crustal extension. This is because at that time the southern San Andreas fault did not exist, and the North American–Pacific plate boundary lay to the east of Baja California. A slight divergence of the Pacific and North American plates, accompanying their sliding motion, caused extension. Crustal extension could have provided an opportunity for magma to reach the surface, explaining the Conejo volcanic series and the volcanic ashes in the Monterey Formation.

An important byproduct of the complex geologic history of the Transverse Ranges is petroleum—most of California's oil and gas reserves lie in this province. Oil and natural gas come from the thermal breakdown of organic matter, mostly microscopic, planktonic marine algae, over geologic time. At lower temperatures, oil is produced, and at higher temperatures, natural gas is produced. The organic matter is usually found finely disseminated in sedimentary rocks, and the end-product—oil or gas—can only be recovered if it is trapped in porous rocks before escaping to the surface. Several things must all occur in the right order to produce recoverable oil reserves. First, source rocks rich in the right kind of organic matter must be present. Second, those rocks have to have been subjected to the right temperatures in their geologic history—warm enough to convert the organic matter to oil or gas, but not too hot to produce only gas. Third, since most organic matter is found in fine-grained rocks like shales that have very low porosity, the oil must migrate to an appropriate reservoir rock with higher porosity—typically a sandstone—if it is to be recoverable. Both oil and gas tend to migrate upward, since they are lighter than water, the other constituent that usually fills pore spaces in rocks. In order to prevent them from escaping to the surface, there must be some kind of trap—an impermeable layer that halts their upward migration.

All of these conditions came together in the southern part of the Transverse Ranges: the Santa Barbara Channel, the Ventura Basin, and the Los Angeles Basin. Here, abundant marine algae was preserved in the Monterey Formation, much of which was deposited in restricted marine basins where low oxygen levels in the water encouraged preservation of the organic matter. Later sedimentation buried these organic-rich sediments. Generally, warmer temperatures are found deeper in the earth's crust, due to the decay of radioactive elements found at depth. Burial of the Monterey Formation provided the right amount of heat to convert the organic matter in it to large amounts of oil and gas. This oil and gas was able to migrate upward into these overlying sediments, some of which were porous sandstones in which petroleum could accumulate. The folded and buckled rocks of the Transverse Ranges provided excellent opportunities to trap these hydrocarbons. Upwarped folds such as the Ventura Anticline allowed oil and gas to migrate upward through porous beds, and overlying impermeable beds trapped much of them below the surface. The folded rocks of the Santa Barbara Channel, the Ventura Basin, and the Los Angeles Basin are the most productive oil reservoirs in California. Nevertheless, some of these traps are less than perfect, and oil and gas do escape to the surface, both offshore and on land. Offshore oil washes up occasionally on beaches from Gaviota to Ventura, forming natural tar balls, which are not to be confused with oil spills resulting from the production and transport of the same oil.

Tarpits Park

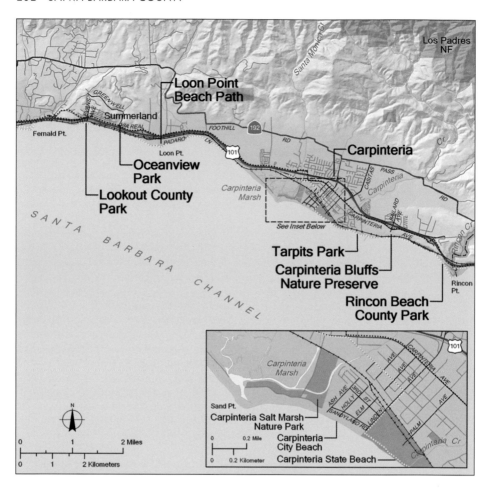

Loon Point
Beach Path

Summerland

GREENWELL

VIA REAL

Fernald Pt.

PADARO LN

Loon Pt.

FOOTHILL

192

RD

101

Carpinteria Marsh

Carpinteria

CASITAS PASS

Carpinteria

RD

BALLARD AVE

LILLIE AVE

CARPINTERIA

AVE

Rincon Cr.

Santa Monica Cr.

Los Padres
NF

Oceanview
Park

Lookout County
Park

See Inset Below

Tarpits Park

Carpinteria Bluffs
Nature Preserve

Rincon Beach
County Park

Rincon
Pt.

SANTA   BARBARA   CHANNEL

N

0   1   2 Miles

0   1   2 Kilometers

Carpinteria Marsh

Sand Pt.

Carpinteria Salt Marsh
Nature Park

Carpinteria
City Beach

Carpinteria State Beach

101

CARPINTERIA AVE

AVE

ASH AVE

HOLLY

ELM ST

LINDEN

SANDYLAND RD

PALM

AVE

Carpinteria Cr.

0   0.2 Mile

0   0.2 Kilometer

Carpinteria State Beach

Summerland to Carpinteria

	Sandy Beach	Rocky Shore	Trail	Visitor Center	Campground	Wildlife Viewing	Fishing or Boating	Facilities for Disabled	Food and Drink	Restrooms	Parking	Fee
Lookout County Park	•							•		•	•	
Oceanview Park											•	
Loon Point Beach Path	•									•	•	
Carpinteria Salt Marsh Nature Park			•			•		•		•	•	
Carpinteria City Beach	•							•		•	•	
Carpinteria State Beach	•	•		•	•		•	•		•	•	•
Carpinteria	•	•	•	•	•	•	•	•	•	•	•	
Tarpits Park	•		•									
Carpinteria Bluffs Nature Preserve			•					•		•	•	
Rincon Beach County Park	•	•			•		•			•	•	

LOOKOUT COUNTY PARK: *End of Evans Ave., Summerland.* This is a small park with a big view of the Santa Barbara Channel. Picnic facilities can be reserved; there are barbecue grills, a playground, volleyball court, outdoor shower, and restrooms. A paved ramp curves down the bluff to the beach, which extends in a wide sweep seaward of the community of Summerland. No lifeguards. From December through April, look for whales from the blufftop. Open 8 AM until sunset. Call: 805-568-2460.

OCEANVIEW PARK: *Via Real at Greenwell Ave., Summerland.* A linear neighborhood park is located inland of Hwy. 101. Use the Padaro Ln. exit from Hwy. 101, then head west on Via Real. The park offers views of the Santa Barbara Channel, picnic tables, and walking paths.

LOON POINT BEACH PATH: *Padaro Ln. at Hwy. 101, Summerland.* A path leads under the road crossing and down the bluff to a long, curving sandy beach. Parking and restroom available, but no park facilities. Call: 805-568-2460.

CARPINTERIA SALT MARSH NATURE PARK: *Ash Ave. and 3rd St., Carpinteria.* The park overlooks part of Carpinteria Marsh. There are interpretive panels and a nature trail; restrooms with running water are at the end

of Ash St., near the beach. The Carpinteria Marsh Nature Park is maintained by the City of Carpinteria; open during daylight hours. Docent-led tours available on Saturdays; call: 805-684-5405. The site is served by the electric Seaside Shuttle, which links downtown Carpinteria with the beach and also connects with buses of the Santa Barbara Metropolitan Transit District; for information, call: 805-683-3702.

The main Carpinteria Marsh is several hundred acres in size and is protected by joint efforts of the Land Trust for Santa Barbara County, the State Coastal Conservancy, the City of Carpinteria, and other entities. Expanded public access into part of the marsh is planned by late 2006, although to protect plants and animals, entry is not allowed into most of it.

CARPINTERIA CITY BEACH: *From the end of Linden Ave. W. to Ash Ave., Carpinteria.* A quarter-mile-long sandy beach extends along the seaward side of Sandyland Rd. Limited parking at the ends of Ash, Holly, Elm, and Linden Avenues. Transit service is provided by the electric Seaside Shuttle; call: 805-683-3702.

CARPINTERIA STATE BEACH: *End of Palm Ave., Carpinteria.* This popular state beach, known as the "safest beach on the coast"

When the hot, dry Santa Ana winds visit the southern California coast, fire is a frequent companion. As the winds rush down canyons and mountainsides, a match, a spark, a cigarette can set the hills ablaze.

The hills do not mind the flames. Chaparral, the brushy vegetation that cloaks coastal mountains in southern California, needs fire to survive. In the dry-summer Mediterranean climate, fire helps dead plants decompose. Fire recycles nutrients, stimulates the germination of seeds, and induces new plants to sprout, bud, and flower. The plants, in turn, welcome fire with leaves and stems that contain volatile oils, have high surface to volume ratios, undergo periodic die-back, and thus help fires burn. The alliance of fire and chaparral has existed for millennia.

Today, urban development fills the lowlands and, increasingly, spreads up the hills into the realm of chaparral and fire. The result is a new fire regime where the area burned contains more suburb and city and less wildland, where blazes are increasingly caused by human error or spite, and where fires must be fought and suppressed.

Fire control techniques may lessen the damage done, but they cannot prevent the perennial reunion of wind, brush, and flame. The most effective fire control methods, such as prescribed burns and creation of firebreaks, mimic the mosaic effect of natural cycles of fire, vegetation regrowth, and maturity. The challenge of those living where the mountains meet the coast is to learn to adapt, as cleverly as the chaparral, to the recurrent presence of fire.

because of the shallow offshore shelf that prevents rip currents, is noted for its good swimming conditions. Lifeguards on duty during the summer. The park also has picnic and day-use parking facilities and a campground sheltered by trees, with 261 campsites, of which 120 have hookups for RVs or trailers; 63 sites are for tents only. Maximum vehicle lengths vary; trailer sanitation station available. Enroute camping allowed. Some campsites are wheelchair accessible. Coin-operated hot showers available. For information or to reserve a beach wheelchair, call: 805-968-1033.

Carpinteria State Beach includes an interpretive display about the Chumash Indians who made this coast their home. The park was once the site of a Chumash village, named Mishopshnow. The village was a center for construction of the wooden plank canoes called *tomols*, used by the Chumash for travel to and from the Channel Islands. When Juan Gaspar de Portolá's overland expedition visited the village in 1769, the Spanish recorded the presence of 38 huts and 300 inhabitants. Portolá's soldiers called the village site *La Carpinteria*, or carpenter shop.

Large deposits of tarry asphaltum lie underneath the beach and cliff area a half-mile east of Carpinteria Creek. Some of the tar seeps are still active. The Chumash used the tar to caulk canoes and waterproof cooking vessels. In the 1920s, fossils were recovered from open tar pits on the beach; they proved to be Pleistocene plants, mammals, birds, insects, and marine invertebrates.

A small riparian woodland of willows and western sycamore is located along the edges of Carpinteria Creek, which bisects Carpinteria State Beach. A sandbar forms a tidal lagoon at the mouth of the creek during the

Chumash *tomol*

summer months. The creek is critical habitat for steelhead trout. At low tide, a small intertidal reef is exposed, providing habitat for a diverse invertebrate community that includes several species of mollusks and crustaceans.

CARPINTERIA: *9 mi. E. of Santa Barbara.* The beach-oriented community of Carpinteria is the urban center of the largely agricultural Carpinteria Valley, an unusually favorable location for growing specialty crops. Rich soils and a long growing season contribute to the area's agricultural importance. There are 3,500 acres of cropland devoted to avocados, lemons, and more unusual fruits such as the sapote and cherimoya. Growers raise lilies, gerbera daisies, and other cut flowers in greenhouses, along with potted plants such as orchids and fuchsias. Although the valley contains less than one percent of Santa Barbara County's agricultural land, it produces one-fifth of the total value of agricultural products.

TARPITS PARK: *E. end of campground access road, Carpinteria State Beach, Carpinteria.* A nine-acre beachfront area with trails and overlooks; Chumash Indian historical site with natural asphaltum seeps. Enter through Carpinteria State Beach. A popular surfing area, called the "tarpits," is located off the south end of the beach.

CARPINTERIA BLUFFS NATURE PRESERVE: *Carpinteria Ave. at Bailard Ave., Carpinteria.* A 52-acre blufftop park overlooks the Santa Barbara Channel and the Channel Islands. The railroad right-of-way separates the Carpinteria Bluffs Nature Preserve from the ocean bluff, and the park does not offer beach access. Grassland on the bluff provides foraging area for raptors, such as the white-tailed kite. Park facilities are simple, in keeping with the natural qualities of the site, and include paths, benches, soccer and baseball fields, and restrooms. Maintained by the City of Carpinteria; call: 805-684-5405. At the base of the bluff is a harbor seal rookery, one of the few remaining in southern California.

RINCON BEACH COUNTY PARK: *Bates Rd., seaward of Hwy. 101, Carpinteria.* A wooden stairway near the parking lot leads down the steep bluff to a sandy beach; there is also a ramp. Picnic facilities, including reservable picnic area, restrooms, and outdoor shower available. The park provides access to Rincon Point, one of the most popular surfing areas along the California coast. Additional parking and beach access is available south of Bates Rd. at Rincon Point in Ventura County. For information or reservations, call: 568-2460.

Unlike a typical high coastal promontory, Rincon Point is a low alluvial fan, composed of cobble material washed down from the nearby mountains. The point is bisected by Rincon Creek, forming the boundary between Santa Barbara and Ventura Counties. The cobble is erosion-resistant and has created a stable landform at the point. Waves refracting off the point create ideal surfing conditions, and Rincon Point is famous well beyond California's borders. Long rides on the waves are possible, from the end of the point in toward Hwy. 101. Can be crowded.

Rincon Beach County Park

Coastal Sage Scrub

THE DENSE THICKETS of shrubs and dwarf trees characteristic of Mediterranean climates with dry hot summers and mild wet winters were generally called chaparral in California. However, botanists have drawn a distinction between "hard chaparral," characterized by large evergreen shrubs with tough waxy leaves, and "soft chaparral," made up of smaller, less dense shrubs with softer leaves. "Chaparral" has come to refer only to hard chaparral, and soft chaparral is now generally referred to as coastal sage scrub because the characteristic species include California sagebrush and white, black, and purple sage. Although the combination of species at any given location varies depending on soils, microclimate, and proximity to the coast, the dominant shrubs that provide most of the physical structure in the habitat will generally be a subset from a fixed list of a dozen or so species. Although the term "coastal sage scrub" is firmly ensconced in the language, some botanists prefer the more generic "coastal scrub" because the sages are sometimes absent.

Like chaparral, coastal sage scrub is a plant community adapted to both drought and fire. Many species produce two types of leaves: larger winter leaves and smaller summer leaves. Other species are drought deciduous, dropping their leaves and going dormant in the summer to avoid water loss. Most species produce seeds that remain viable for long periods and are stimulated to sprout by heat or the chemicals found in ash, and many of the shrubs have the ability to resprout from root crowns. During the first spring or two following a fire, the coastal sage scrub community will be awash in color from the many wildflowers that have been induced to sprout and that can then take advantage of the temporary lack of shading by larger plants. However, within a few more years the shrubs will again dominate the landscape, and the only evidence of the fire will be the charred stems of the woodier species.

Although adapted to fire, coastal sage scrub is not adapted to frequent fires. Where human presence increases ignition sources, frequent fires result in reduced native biodiversity, increased abundance of exotic annual grasses that produce huge amounts of seed, and domination by a few strong resprouters, like laurel sumac, that can cope with the new fire regime. Doubling the fire frequency from once every 24 years to once every dozen years may completely change the plant community and its associated wildlife.

Coastal sage scrub plant community

Black sage (*Salvia mellifera*) is a true sage and is a member of the *Lamiaceae*, or mint, family. You can smell the aromatic oils of black sage as you walk by without even touching the plant. Like most members of this family, black sage has square stems. It has dainty, light-blue-to-purple flowers that grow in compact whorls spaced two to three inches apart on the main stems. Hummingbirds, butterflies, and other small insects love to gather nectar from the fragrant flowers that bloom between March and July. Honeybees utilize this plant in the production of honey. The common name, black sage, was given to this species because the whorls of blooms remain on the stems after they set seed, forming dark spheres along the dry stalks.

Black sage

Coyote brush (*Baccharis pilularis*) is a pioneer plant in communities such as coastal sage scrub and chaparral. When native vegetation is removed from an area by bulldozer, tilling, or grazing and trampling animals, this is one of the first woody native plants to advance into the idle cleared land. It has been used extensively in restoration projects due to its ability to survive under these conditions. Coyote brush blooms in the fall, and it is dioecious; this means that each bush is either a male or a female. The males, which are pollen bearing, produce an intoxicating scent in fall right before the rains come. The female bushes, which are seed bearing, look fuzzy in the fall when their seeds are ready to be blown away upon the winds.

Coyote brush

California sagebrush (*Artemisia californica*) has fragrant, feathery, gray foliage and grows three to four feet tall. The genus name *Artemesia* is from the Greek goddess Artemis, who is said to have benefited so much from a plant of this family that she gave it her own name. The Cahuilla Indians were known to use the leaves from California sage to relieve colds. California sage is not a true sage; it is a member of the *Asteraceae*, or sunflower, family. The small, green, sunflower-like blossoms open in the fall between August and December. If you spy California sage during your adventures in coastal sage scrub, take a moment to rub a few leaves between your fingers to smell its lovely scent.

California sagebrush

Poison oak

Poison oak (*Toxicodendron diversiloba*) is not an oak, but rather a member of the sumac family. It grows as a shrub or occasionally as a vine, with shiny leaves composed of three coarsely lobed leaflets, which gave rise to the old adage, "leaflets three, let them be." Poison oak is one of the most common causes of contact dermatitis in North America. A characteristic rash results when one's skin brushes against the plant. This is a true allergic reaction, and itching can be quite intense. Some people appear to be tolerant to poison oak, but the mechanisms for this are not well understood.

Morro shoulderband snail

The **Morro shoulderband snail** (*Helminthoglypta walkeriana*) is a small land snail, about an inch in diameter, that occurs in coastal sage scrub and dune communities in western San Luis Obispo County. It was listed as endangered by the U.S. Department of Interior in 1994. This species can be identified by the one narrow, dark brown spiral band on its somewhat translucent shell. The Morro shoulderband snail was first described in 1911 as living in or near the sandy soils of San Luis Obispo County south of Cayucos. Since then its range has diminished due to habitat destruction and degradation. Like most snails, the Morro shoulderband snail is most active at night, reproduces in the rainy season, and estivates during the dry season. Little else is known about this elusive snail's life history.

Coast horned lizard

The **coast horned lizard** (*Phrynosoma coronatum*) has several unique adaptations to deter predators. First, the lizard is able to change its color to match the surrounding environment, a mechanism known as cryptic coloration. It blends in so well to the background that it is usually only visible when it moves. Another adaptation of the coast horned lizard is that, when threatened, it is capable of shooting a small stream of blood from its eyes. When the lizard is in danger the blood pressure in its head increases, and tiny blood vessels in the corner of the eyes rupture and blood shoots out. This may provide a momentary distraction and allow the lizard to move to safety. Ants are the favorite food of horned lizards. However, non-native ants have displaced or eradicated the native ant species upon which the coast horned lizard preys.

The **white-crowned sparrow** (*Zonotrichia leucophrys*) has a pink bill, and its head has two distinct black stripes with a broad white crown stripe between them. Much of what we know about bird song development and geographical variation in song has been gained from studying the white-crowned sparrow. A young male white-crowned sparrow learns the song of its region by listening to the birds in its neighborhood. Different neighborhoods of white-crowned sparrows have different "dialects." Because the sparrows learn the songs from a specific region, they generally return to the same area where they were raised. Interestingly, males raised on the edge of two regions or neighborhoods may be bilingual and able to sing both dialects.

White-crowned sparrow

Allen's hummingbird (*Selasphorus sasin*) is a small, compact hummingbird with a glittering green crown, a white breast, and rufous sides. The male's throat, or gorget, is iridescent coppery-red. Allen's hummingbirds have one of the smallest breeding ranges of all U.S. hummingbirds. They breed in a narrow strip along the Pacific coast from southwest Oregon to southern California. These perky little hummingbirds arrive on the coast in the early spring, and in very short order, the males establish territories and attract females using extravagant aerial displays accompanied by vocalizations. The males maintain territories that overlook open coastal scrub by perching in conspicuous places and engaging in territorial battles with their intruders.

Allen's hummingbird

Coastal sage scrub is home to dozens of species of butterflies, many of which, as larvae, are specialized herbivores on one or a few species of plants. The **common checkerspot butterfly** (*Euphydryas chalcedona*) is found in large populations in coastal foothills from Oregon to Baja California. In scrub habitats, checkerspots lay their eggs on sticky monkey flower and on the bee plant (*Scrophularia*) and the species' activity patterns are closely linked to the patterns of growth and dormancy of their host plants. Caterpillars hatch and begin feeding around June but soon enter a quiescent "diapause" during the summer drought. Caterpillars become active again following the onset of winter rains and continue feeding until they pupate in the spring. Look for the strikingly patterned adults from April to July.

Common checkerspot butterfly

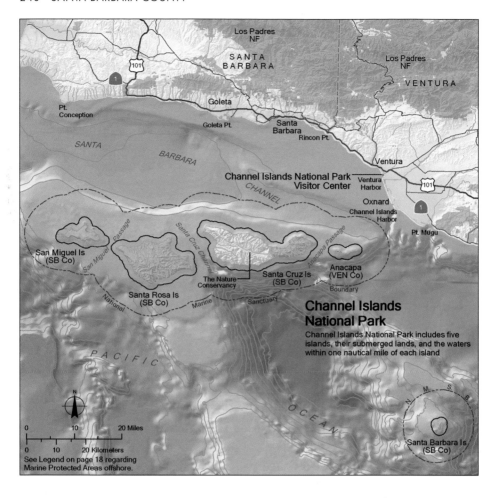

Los Padres
NF

SANTA
BARBARA

Los Padres
NF

VENTURA

Goleta

Pt.
Conception

Goleta Pt.

Santa
Barbara

Rincon Pt.

Ventura

SANTA

BARBARA

Channel Islands National Park
Visitor Center

Ventura
Harbor

CHANNEL

Oxnard

Channel Islands
Harbor

Santa Cruz Channel

Santa Cruz Passage

Anacapa Passage

Pt. Mugu

San Miguel Passage

San Miguel Is
(SB Co)

The Nature
Conservancy

Santa Cruz Is
(SB Co)

Anacapa
(VEN Co)

National

Santa Rosa Is
(SB Co)

Marine

Sanctuary

Boundary

PACIFIC

Channel Islands National Park

Channel Islands National Park includes five
islands, their submerged lands, and the waters
within one nautical mile of each island

OCEAN

N

0   10   20 Miles

0   10   20 Kilometers
See Legend on page 18 regarding
Marine Protected Areas offshore.

N. M. S. B.

Santa Barbara Is
(SB Co)

Anacapa Island historic structures

Channel Islands National Park

	Sandy Beach	Rocky Shore	Trail	Visitor Center	Campground	Wildlife Viewing	Fishing or Boating	Facilities for Disabled	Food and Drink	Restrooms	Parking	Fee
Cuyler Harbor Beach (San Miguel Island)	•		•		•	•				•		
Campground (San Miguel Island)			•	•	•	•				•		•
Water Canyon Beach (Santa Rosa Island)	•		•	•		•	•			•		
Campground (Santa Rosa Island)			•		•	•				•		•
Scorpion Beach (Santa Cruz Island)	•	•	•	•	•	•	•	•		•		
Prisoners Harbor (Santa Cruz Island)			•	•		•	•			•		
Smugglers Cove (Santa Cruz Island)			•	•	•					•		
Landing Cove (Anacapa Island)			•	•		•	•			•		
Campground (Anacapa Island)			•		•	•				•		•
Landing Cove (Santa Barbara Island)				•		•	•			•		
Campground (Santa Barbara Island)			•	•	•	•				•		•

CUYLER HARBOR BEACH: *N.E. side of San Miguel Island.* A two-mile-long white sand beach greets visitors to the westernmost Channel Island. The beach and the mile-long trail to the campground and ranger station are the only parts of the island that visitors can explore without a ranger. Check with the National Park Service if you want to go beyond those points. Visitors willing to brave the sometimes harsh weather conditions will be rewarded with unique geological and ecological sights. The five-mile-roundtrip trail to the caliche forest takes hikers to a collection of sand-castings of ancient vegetation. The island is also home to a recovering seal and sea lion breeding ground; a 16-mile out-and-back hike to Point Bennett offers visitors the chance, at certain times of the year, to see up to 30,000 animals of up to five pinniped species.

CAMPGROUND: *1 mi. S. of Cuyler Harbor, San Miguel Island.* A steep trail from the landing leads one mile to a primitive campground. There are nine campsites, each with a windscreen. Pit toilets only; bring your own water.

WATER CANYON BEACH: *N.E. side, Santa Rosa Island.* Extending two miles southeast from the Santa Rosa Pier is pristine Water Canyon Beach. An option for day visitors is the new three-mile Cherry Canyon Loop Trail, which winds through hillsides dotted with endemic plants and scrub oaks to a headland overlooking bright blue water. The five-mile loop to the Torrey Pines grove requires moderate effort; the subspecies of Torrey Pines found on Santa Rosa Island is one of the rarest pines in the world. Visitors should be prepared for windy conditions. A restroom with running water is located inland from the pier, near the Vail and Vickers Ranch House. Archaeological resources on the island include the 13,500-year-old remains of a woman, believed to be the oldest human remains found anywhere in North America.

CAMPGROUND: *1.5 mi. S. of the landing, Santa Rosa Island.* From the pier, the campground is accessible by hiking along a flat dirt road. Each of the 15 campsites has a windscreen. Restrooms and showers are a recent addition. Visitors can explore trails, take a short stroll to a white sand beach, or view the plant life and wildlife of the island. Endemic plants include the island bush poppy, Indian paintbrush, and California holly. A captive breeding program on the island is underway, with the goal of reintroducing the endangered island fox to its natural habitat.

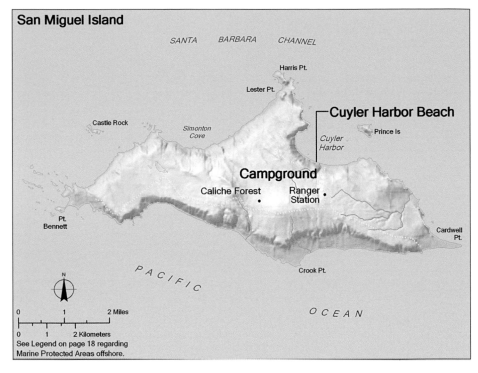

San Miguel Island

SANTA   BARBARA   CHANNEL

Harris Pt.

Lester Pt.

Cuyler Harbor Beach

Castle Rock

Simonton Cove

Prince Is

Cuyler Harbor

Campground

Caliche Forest   Ranger Station

Pt. Bennett

Cardwell Pt.

PACIFIC

Crook Pt.

OCEAN

N

0   1   2 Miles

0   1   2 Kilometers

See Legend on page 18 regarding Marine Protected Areas offshore.

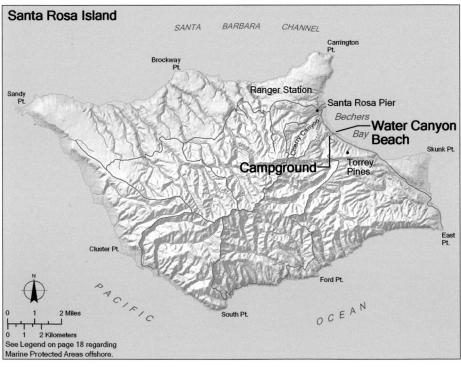

Santa Rosa Island

SANTA   BARBARA   CHANNEL

Carrington Pt.

Brockway Pt.

Ranger Station

Sandy Pt.

Santa Rosa Pier

Bechers

Water Canyon Beach

Cherry Canyon

Bay

Skunk Pt.

Campground   Torrey Pines

East Pt.

Cluster Pt.

Ford Pt.

PACIFIC

South Pt.

OCEAN

N

0   1   2 Miles

0   1   2 Kilometers

See Legend on page 18 regarding Marine Protected Areas offshore.

Channel Islands National Park includes San Miguel, Santa Rosa, Anacapa, and Santa Barbara Islands and the eastern portion of Santa Cruz Island, including one nautical mile offshore. Because of the Channel Islands' remote location and sparse facilities, visitors must plan ahead. Camping reservations for all islands are required; call: 1-800-365-2267. Virtually everything visitors will want must be brought along; water is available at only two locations, Scorpion Beach campground on Santa Cruz Island, and the Santa Rosa Island campground. Both day visitors and campers must bring food and water and should familiarize themselves with facilities available at their destination. Visitors to San Miguel Island land by skiff on the beach and should dress for getting wet. Boat visitors to the other islands must climb up steel-rung ladders at the landings. Surfing is possible at various remote points around the islands; surfing sites are best reached by private boat.

Island Packers, Inc., based in Ventura, offers boat trips to all islands. Trips to Anacapa, Santa Cruz, and Santa Rosa Islands are offered year-round, weather permitting, while trips to San Miguel and Santa Barbara Islands are available from spring until fall. Island Packers offers day trips and also transports camping, kayaking, and diving gear for campers; for information, call: 805-642-1393. Boat reservations are strongly recommended at least two weeks in advance, especially in summer and for visitors transporting gear.

Island Packers does not offer kayak rentals, but several outfitters will bring kayaks to the boat prior to departure.

Kayakers should be prepared for rough, windy ocean conditions. Santa Cruz and Anacapa Islands are the safer options for kayaking, but can still present danger. The National Park Service and island concessionaires recommend that only highly experienced kayakers embark on trips off Santa Rosa and San Miguel Islands. Winds at the Channel Islands normally blow from north to south, and kayakers are warned that a southbound trip might entail a tough return against the wind.

Santa Barbara–based Truth Aquatics offers a variety of full-service trips to the islands, including diving trips; call: 805-963-3564. Channel Islands Aviation offers public air transportation year-round to Santa Rosa Island, which has a small landing strip; call 805-987-1301.

Kayaker, Anacapa Island

Santa Cruz Island

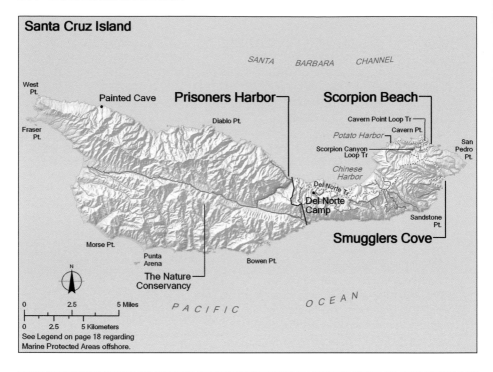

SANTA   BARBARA   CHANNEL

West
Pt.

Painted Cave   **Prisoners Harbor**   **Scorpion Beach**

Diablo Pt.   Cavern Point Loop Tr

Fraser
Pt.   Cavern Pt.

Potato Harbor

San
Pedro
Pt.

Scorpion Canyon
Loop Tr

*Chinese
Harbor*

Del Norte Tr

Del Norte
Camp

Sandstone
Pt.

Morse Pt.   **Smugglers Cove**

Punta
Arena

Bowen Pt.

The Nature
Conservancy

N

0   2.5   5 Miles   *P A C I F I C*   *O C E A N*

0   2.5   5 Kilometers
See Legend on page 18 regarding
Marine Protected Areas offshore.

Anacapa Island

SANTA   BARBARA   CHANNEL   Anacapa Lighthouse

Landing Cove

West Anacapa   Cathedral Cove Tr   Arch
                                                                  Rock
Rat Rock   *Cathedral Cove*

Inspiration Point Tr   **East Anacapa**

*Frenchys
Cove*   Middle Anacapa   Overlook Tr

N

0   0.5   1 Mile   Cat Rock   **Campground**

0   0.5   1 Kilometer   *P A C I F I C*   *O C E A N*
See Legend on page 18 regarding
Marine Protected Areas offshore.

Santa Barbara Island

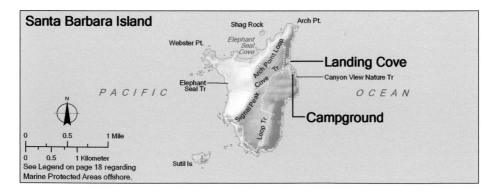

Shag Rock   Arch Pt.

Webster Pt.   *Elephant
Seal
Cove*

Landing Cove

Arch Point Loop

Canyon View Nature Tr

Elephant
Seal Tr

P A C I F I C   Signal Peak   Cove Tr   *O C E A N*

N

Loop Tr   **Campground**

0   0.5   1 Mile

0   0.5   1 Kilometer
See Legend on page 18 regarding   Sutil Is
Marine Protected Areas offshore.

SCORPION BEACH: *N.E. end of Santa Cruz Island.* Scorpion Beach is the main landing for Santa Cruz Island, the largest of the Channel Islands. The island's 96 square miles encompass rugged mountain ranges, quiet canyons, and seasonal creeks out of sight of the sea, where a visitor might forget the island setting. Visitors can kayak, snorkel, dive, or swim in the bright blue ocean water; sunbathing is possible at protected, sand-and-cobblestone Scorpion Beach. The beach's proximity to majestic sea caves and rocks, relatively calm and clear water, and huge kelp beds make it a popular place for water-oriented adventures. Picnic tables are located at the beach, and pit toilets are nearby.

From Scorpion Beach, hikers have access to a network of trails of varying length and difficulty. The two-mile Cavern Point Loop Trail offers day visitors a breathtaking coastal view. The trail rises steeply to Cavern Point, high above the sea; on a clear day, mountains on the mainland can be seen. Another two miles of hiking leads to Potato Harbor and its sheer island cliffs, a great trek for an ambitious day-trip visitor. The Scorpion Canyon Loop, four-and-a-half miles long, winds up steep inland ridges where the island scrub jay, found only on Santa Cruz Island, may be spotted. The more strenuous eight-mile Montañon Ridge Trail takes hikers deep into the island's interior, traversing steep ridges dotted with scrub oak. Maps of all hiking trails are located at the information board at Scorpion Beach.

The largest campground on any of the islands is located one-quarter mile west of Scorpion Beach in a quiet eucalyptus grove. Many campers walk to Potato Harbor to take in the sunset before returning to the campground. There are 40 campsites with picnic tables and food lockers. Fires are prohibited; campers should bring their own stoves. Running water and pit toilets are available.

PRISONERS HARBOR: *N. side of Santa Cruz Island.* A cobblestone beach, hiking trails, and picnic tables can be reached from the harbor, which is visited regularly by Island Packers. Pit toilets available. A three-and-a-half-mile trail leads to Del Norte Camp, with four campsites; no running water. Prisoners Harbor and Scorpion Beach are linked by trail, making possible a long backpack-

Moonset, Santa Cruz Island

Santa Cruz Island

ing trip from one landing to the other. Prisoners Harbor is located at the edge of the National Park Service–administered part of Santa Cruz Island. Only a limited number of guided hikes are offered on the part of the island managed by the Nature Conservancy.

SMUGGLERS COVE: *3.5 mi. S. of Scorpion Beach, Santa Cruz Island.* On the more protected side of the island, the cove offers access to a cobblestone beach, a good place for fishing. The island concessionaires do not land here, and a visit requires either a strenuous seven-mile roundtrip hike or a lengthy kayak trip from Scorpion Beach. Pit toilets only; no camping.

LANDING COVE: *N.E. end of East Anacapa Island.* Bounded by high, sheer volcanic cliffs, the island is difficult for boat landings; Landing Cove is one of the only places to gain access to Anacapa Island. There are actually three small isles: West, Middle, and East Anacapa, the only one with any facilities. Landing Cove has a pulley to take kayaks in and out of the water and is a popular location for divers. For non-divers, the National Park Service offers the Kelp Video Program, in which a diver with an underwater video camera provides a look at underwater life in Landing Cove, while visitors view the show at the landing. Expect to see a diverse array of sea life, including fishes such as the senorita and sheephead, spiny lobsters, sea stars, and perhaps a sea lion in the background. The underwater video program is offered every Tuesday and Thursday at 2 PM, from Memorial Day to Labor Day.

From the landing, a long stairway leads up to the ranger station, information board, and a small visitor center; picnic tables and pit toilets are available. Sights include the Anacapa Lighthouse, the last permanent lighthouse built on the west coast. East Anacapa is a small island and only short hikes are available. From May through July, Anacapa is home to a thriving western gull breeding colony, which dominates the sights and sounds of the somewhat barren island. The small size of the island makes Anacapa (actually part of Ventura County) a good choice for day visitors seeking an introduction to the Channel Islands.

CAMPGROUND: *.5 mi. W. of Landing Cove, East Anacapa Island.* There are seven campsites, each with picnic table and food locker.

There are pit toilets, but no water here or elsewhere on the island. The site offers little shelter from the elements; camping is most popular with divers or kayakers who plan to embark at Landing Cove.

LANDING COVE: *N.E. end of Santa Barbara Island.* Only one square mile in size, Santa Barbara Island has a five-mile network of trails. Visitors can snorkel or dive near the landing cove; there are no beaches on the steeply sided island. The spring months offer beautiful wildflower displays. Santa Barbara's endemic plant and animal populations are slowly recovering from years of ranching. Species to look for include the island night lizard and cliff-side breeding colonies of Xantus's murrelets; ten other species of seabirds nest on the island.

CAMPGROUND: *.5 mi. S. of Landing Cove, Santa Barbara Island.* Located uphill from the landing, the campground has eight sites. No water is available.

Painted Cave, Santa Cruz Island

Sea caves are common along the rocky California coast. Nowhere, though, are they more numerous and better developed than along the margins of the Northern Channel Islands. Sea caves range from large rooms with high ceilings and shimmering light to dark, narrow tunnels that stretch underground for hundreds of feet. Anacapa Island alone hosts more than 130 caves. Santa Cruz Island is home to the largest known sea cave in the world. Known as Painted Cave because of its colorful rock types, lichens, and algae, the ceiling is nearly 160 feet above the water at the entrance. Nearly a quarter-mile long and 100 feet wide, the cave sports a spring waterfall cascading over its entrance. Sea caves have been eroded into the volcanic rock on Santa Cruz Island where pounding surf has attacked zones of weakness—typically cracks or faults that split the rock. Most sea caves can only be entered by kayak or boat. Kayakers should have knowledge of safety techniques for sea cave exploration; helmets should be worn at all times. Less experienced visitors should go with a tour guide; several commercial outfitters conduct tours of the caves.

Northern Channel Islands

A T THE PEAK of the last ice age about 18,000 years ago, a single large island, now dubbed Santarosae, lay off the Santa Barbara coast. As the ice melted, the sea gradually rose over three hundred feet, submerging three-quarters of the island. The highest ridges remained exposed but now separated from one another by an ocean moat. The exposed peaks of Santarosae are the present-day northern Channel Islands: San Miguel, Santa Rosa, Santa Cruz, and Anacapa Islands. This island chain, thought to be an extension of the Santa Monica Mountains, has apparently never been connected to the mainland by a land bridge. Therefore, all the plants and animals on these islands had to cross miles of ocean in order to colonize.

Successful colonization across an ocean barrier is a relatively rare event for many kinds of species, and most that arrived had few opportunities to interbreed with individuals from mainland populations. This genetic isolation has resulted in a relatively high degree of endemism on the islands. Endemic species or subspecies are those that are only found in a particular region; in this case, only on one or more of the northern Channel Islands. For example, each island has its own species of deer mouse; nine species of the drought-tolerant, succulent plant, *Dudleya*, are found only on the islands. On the other hand, the islands are home to only one endemic bird species, the island scrub jay, probably a reflection of the greater colonizing ability of flying species. Dwarfism is a common phenomenon on islands, and cat-sized island foxes can be seen today, with a different subspecies on each Channel Island on which they occur.

Humans may have visited the islands as long ago as 13,000 years, and the Chumash people established settlements by about 7,000 years ago. Although the Chumash no doubt affected the ecology of the islands, it was the arrival of Europeans that doomed both the island Chumash and their natural world. Although first visited in 1542 by Cabrillo, the Channel Islands were spared colonial devastation for 200 years. However, the influx of Europeans and Americans during the 50 years following the arrival in California of Spanish missionaries in 1769 forever changed the face of the islands. The Chumash died of exotic diseases or were removed to the mainland; fur traders drove sea otters, fur seals, and elephant seals to local extinction; goats, pigs, and sheep were introduced to the islands where, to this day, they have had effects on biological communities ranging from disruption to devastation; and in the mid-1800s, settlers moved to the islands to farm and ranch.

Resource protection gained ground when Anacapa Island became part of the Channel Islands National Monument in 1938, the Nature Conservancy took over protection of Santa Cruz Island in 1978, and, in 1980, the Channel Islands National Park was established. Through a series of acquisitions and donations, by 1997 all the northern Channel Islands were being managed for resource protection and restoration. One of the biggest remaining threats is the continued presence of feral goats, pigs, and sheep. However, an aggressive program of removal coupled with hunting may yet result in the extirpation of these reminders of a period when exploitation overwhelmed stewardship. The National Park Service and the Nature Conservancy are making great strides in protecting and restoring the island habitats and providing opportunities for visitors to experience these magical places.

The **island bush poppy** (*Dendromecon harfordii*) is a large, evergreen, flowering shrub endemic to the chaparral community of the Channel Islands. The island bush poppy produces beautiful golden flowers in spring and summer. Each flower has four large bright yellow petals. The flowers are solitary and are terminal on the ends of short branches. The island bush poppy is related to the mainland bush poppy (*Dendromecon rigida*), however these two bush poppies are distinct species. The physical differences between the two are subtle, but due to the different environmental conditions and lack of gene flow between them, they have evolved into different species.

Island bush poppy

The **island fox** (*Urocyon littoralis*) is the smallest fox species in the United States and the largest mammal endemic to the Channel Islands. The island fox has markings that are similar to those of the related gray fox, but the island fox is only about the size of a house cat. The island fox lives on six of the eight northern and southern Channel Islands. Each island has its own distinct subspecies of the fox. The Chumash Indians apparently kept the island fox as pets more than three thousand years ago and probably helped to spread their populations among the southern Channel Islands. Due to serious population declines, four out of the six subspecies of island fox have been listed as endangered pursuant to the federal Endangered Species Act.

Island fox

The **island scrub jay** (*Aphelocoma insularis*) is a distinct species, even though it looks very similar to the western scrub jay. The island scrub jay has a stouter bill, larger overall size, and the plumage is a brighter blue. Its range is restricted to Santa Cruz Island, the largest and most topographically diverse of the Channel Islands. Why are the mainland and the island scrub jay different? When an animal colonizes an island, the new environmental conditions, the absence of predators or competitors, and the limited gene pool virtually guarantee that changes will occur. Adaptations that were important on the mainland environment may become less important; new adaptations evolve to address the new environmental conditions.

Island scrub jay

California opaleye

California opaleye (*Girella nigricans*) are characterized by an overall dark olive green coloration with one or two light spots along the back at the base of the dorsal fin; the eyes are a beautiful opalescent blue-green. Opaleye grow to about 25 inches and weigh an average of about ten pounds. They prefer shallow water and rocky areas. In the spring they can be found in large schools in shallow reef areas that have an abundance of algae growth. Opaleye are omnivorous, meaning they eat both animals and plants.

California spiny lobster

The **California spiny lobster** (*Panulirus interruptus*) lacks the large pinching claws of its Atlantic relatives. The California spiny lobsters spend the day hidden in rocky caves and crevices and emerge at night to scavenge the reef for sea urchins, mussels, and worms. During winter months they are found offshore at depths of 50 feet or more. In late March through May, lobsters move into warmer nearshore waters with a depth of less than 30 feet. In late October or early November, dropping water temperatures and storm surge will encourage the lobsters to move offshore again. Lobsters have unusual eyes, located at the end of movable stalks, similar to the periscope of a submarine. Like a bee's eyes, lobster eyes are compound, so the lobster sees a mosaic of many images. While the image may be a bit blurry compared to what we see, the lobster's eyes make it especially skillful at detecting movement in its environment.

Senorita fish

The cigar-shaped **senorita fish** (*Oxyjulis californica*) is mostly yellow with a black bar on its tail. It is a type of cleaner fish. These tidy little fish pick small parasites off larger fish. This arrangement not only benefits the senorita, which receives a meal, but also the other fish, which rid themselves of undesirable parasites. The senorita is found in kelp forests and reefs along the coast from northern California to central Baja California. Senorita fish are active during the day and sleep at night. They have an interesting bedtime routine; after sunset, they swim headfirst into the sand, burying themselves for the night. Before sunrise, they shake themselves loose from the sand and head out for another day.

Channel Islands National Marine Sanctuary

THE CHANNEL ISLANDS National Marine Sanctuary is the southernmost of four national marine sanctuaries off California's coast. The Marine Sanctuary includes the waters extending a distance of six nautical miles from the mean high tide line along the shores of San Miguel, Santa Rosa, Santa Cruz, Anacapa, and Santa Barbara Islands. These are the same five islands included within the boundaries of the Channel Islands National Park; protection of the marine waters around these islands is coordinated with management of land areas by the National Park Service and the Nature Conservancy, which manages most of Santa Cruz Island.

The Marine Sanctuary is located at the intersection of cold currents flowing south along the coast, and warmer waters moving north. The biological diversity that results from this mixing of environments is extraordinary. More types of marine mammals are found here than anywhere else in the world in an area of comparable size; there are more than 27 species of whales and dolphins, including the blue whale, largest of all whales (and of all living creatures). In recent years, blue whales have been sighted in the Marine Sanctuary in increasing numbers. Also found in the area are five species of seals and sea lions, four of which breed on the northern Channel Islands. More than 25 species of sharks are found around the islands, including one of the world's largest fishes, the basking shark, which grows to 45 feet long.

Habitats within the Channel Islands National Marine Sanctuary include huge kelp forests that harbor fish and invertebrates, rocky shores and reefs, sandy beaches, seagrass meadows, and open ocean. Other resources protected by the Marine Sanctuary include over 100 historic shipwrecks and artifacts of the Chumash indigenous people. Research supported by staff of the sanctuary addresses a wide range of topics. One project has addressed the movements of giant sea bass, white sea bass, kelp bass, and California sheephead, both within and outside the Anacapa Island State Marine Reserve. Tracking is accomplished by acoustic transmitters that are surgically implanted in the fish; data on the movement of the fish are collected by receivers on the sea floor. Learning more about the movements of these species will help in designing fishing regulations around the islands.

Blue whale tail

Humpback whale and shearwater

The California Department of Fish and Game provides additional protection to marine resources within the Channel Islands National Marine Sanctuary through designation of 12 marine protected areas (MPAs). Ten of these areas allow no fishing, in order to provide "safe zones" for species that are experiencing declining numbers or pressure due to overfishing. One marine protected area, located off Santa Cruz Island, allows only recreational fishing for finfish and lobster, and another, located off Anacapa Island, allows recreational fishing and limited commercial fishing for lobster. Detailed descriptions of the areas and rules for the marine protected areas are available from the Department of Fish and Game; call: 805-568-1231, or see www.dfg.ca.gov.

Anacapa Island was the site in 1992 of the first Great Annual Fish Count, an event designed to utilize the efforts of volunteer divers and snorkelers to monitor the populations of ocean fish. The count was modeled on the Audubon Christmas Bird Count, an annual bird census conducted by volunteers. The Marine Sanctuary and the Reef Environmental Education Foundation, or REEF, deploy divers several times a year to conduct fish counts around the islands. The Marine Sanctuary holds fish identification training sessions for volunteer divers in advance of the counts. Information from periodic counts such as these is useful to scientists for monitoring population trends of key species, as well as to wildlife viewers who are interested in finding "hotspots."

The Marine Sanctuary hosts workshops and lectures for teachers, students, and the public about resource protection, monitoring, and research. The Shore to Sea lecture series presents guest researchers who discuss topics that include marine mammal studies, seabird research, archaeology, and maritime heritage. The efforts of volunteers drive important programs of the Marine Sanctuary. The Channel Islands Naturalist Corps is made up of volunteer naturalists who accompany visitors on hikes, whale-watching trips, and the concessionaire vessels that carry visitors to the islands. The Naturalist Corps is a ready source of information for visitors. Volunteers also serve on the Sanctuary Advisory Council, which is a mechanism for interested persons in getting involved with the Marine Sanctuary. To learn more about the Channel Islands National Marine Sanctuary, see: www.channelislands.noaa.gov, or call the sanctuary's Santa Barbara office at 805-966-7107 or Ventura office at 805-382-6149 ext.102.

Page opposite: Faria County Park, Ventura County

Ventura County

Ventura County

VENTURA COUNTY'S distinctive coastline is divided into three segments. The northern and southern sections of coast are bounded by mountains of the Transverse Ranges that rise abruptly from the sea, leaving a narrow shelf along the water's edge. The steep slopes of the Santa Ynez Mountains form the county's northern coast, while the Santa Monica Mountains rise above the south coast. Sandy beaches border much of the northern shoreline, although the location of the coastal highway and nearby railroad tracks shape public access to the shore. No freeway exits are located on the six-mile stretch of Hwy. 101 northwest of Ventura; beach-goers should take the old Pacific Coast Hwy. instead of the freeway. Parking restrictions along the old highway, where it borders the beach, are well posted and well enforced. South of Point Mugu, Pacific Coast Hwy. hugs the shoreline. The highway is characterized on one side by sandy beaches separated by rock outcroppings and on the other by the steep slopes and canyons of the Santa Monica Mountains. Both sides of the road offer recreational opportunities.

Between the city of San Buenaventura and Point Mugu, the middle reach of the county's coastline is characterized by a series of wide sandy beaches, broken only by the mouths of two harbors and two rivers, plus smaller streams. The flat Oxnard Plain, with its rich soil, borders much of this coast, with the cities of Oxnard and Port Hueneme occupying part of the plain. Ventura Harbor and Channel Islands Harbor provide boating, sailing, and other water-oriented recreation. Extensive wetlands exist where the Ventura and Santa Clara Rivers reach the sea; Mugu Lagoon at the south end of the Oxnard Plain is one of the largest remaining salt marshes in southern California, but it lies within the Naval Air Station Point Mugu and is off-limits to visitors. Splendid white sandy beaches, some of the region's widest, are found in this middle part of Ventura County's coastline. Mild weather year-round and ocean water temperatures that are warmer than elsewhere on the Central Coast contribute to the appeal of the shoreline.

Ventura County's mix of sheltering mountains and south-facing seashore produces an enviable environment not only for visitors, but also for agriculture. One of California's leading agricultural counties, Ventura County produces famous crops of sweet, juicy strawberries, the county's leading crop in value. Lemons, avocados, celery, tomatoes, raspberries, peppers, and Valencia oranges also rank among the area's top ten crops. Cut flowers and nursery stock, including orchids and poinsettias, are also multi-million dollar commodities. In all, the county's agricultural products were worth $1.2 billion dollars in 2005.

South Coast Area Transit serves the Ventura, Oxnard, and Port Hueneme areas; call: 805-487-4222.

For visitor information, contact:

Ventura Visitors and Convention Bureau, 1-800-483-6214, or 805-648-2075.

City of Ventura Chamber of Commerce, 805-676-7500.

Oxnard Convention and Visitors Bureau, 1-800-269-6273.

Oxnard Chamber of Commerce, 805-983-6118.

Port Hueneme Chamber of Commerce, 805-488-2023.

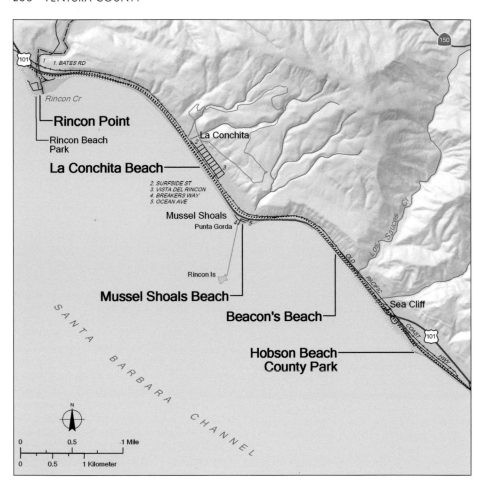

Hobson Beach County Park

Ventura County North

	Sandy Beach	Rocky Shore	Trail	Visitor Center	Campground	Wildlife Viewing	Fishing or Boating	Facilities for Disabled	Food and Drink	Restrooms	Parking	Fee
Rincon Point	•	•								•	•	
La Conchita Beach	•											
Mussel Shoals Beach	•	•									•	
Beacon's Beach	•										•	
Hobson Beach County Park	•	•			•	•	•	•	•	•	•	•

RINCON POINT: *Off Hwy. 101, S. of Bates Rd.* Exit Hwy. 101 at Bates Rd. and turn seaward to two beach accessways: to the right, Santa Barbara County's Rincon Beach Park, and to the left, a public parking lot in Ventura County with a short path leading to a mixed cobble and sand beach. Rincon Point is a major surfing spot. Private property is adjacent; do not trespass.

LA CONCHITA BEACH: *Along Hwy. 101 between Rincon Point and Mussel Shoals.* A sandy beach is just north of Mussel Shoals. Shoreline access is across the riprap revetment along the Hwy. 101 shoulder, except where parking is prohibited by "Emergency Parking Only" signs. Use caution, as traffic moves fast here; no parking or pedestrians allowed along the freeway shoulder to the north.

MUSSEL SHOALS BEACH: *W. of old Pacific Coast Hwy., Mussel Shoals.* There is a sandy beach north of the point called Punta Gorda and the residential community of Mussel Shoals; a rocky beach with tidepools lies to the south. Rincon Island is an artificial island located offshore Punta Gorda and connected to land by a private causeway. The island was built in 1958 for oil extraction.

Access to the public beach is at the far south end of Mussel Shoals, past the hotel, and at the west end of Ocean Ave. Limited parking is available along old Pacific Coast Hwy.; turn right as you exit Hwy. 101. No public parking in the private residential area.

BEACON'S BEACH: *N. end of old Pacific Coast Hwy., S. of Mussel Shoals.* A long sandy beach is located adjacent to the former Mobil Oil piers, north of the residential community of Sea Cliff. Two beach accessways begin on the inland side of the Hwy. 101 freeway, along old Pacific Coast Hwy., now a dead end road; use the Sea Cliff freeway exit. One accessway is a vehicle underpass beneath Hwy. 101; parking is allowed on the inland side of the shoreline frontage road, except from 10 PM to 5 AM. A second underpass beneath Hwy. 101, for pedestrians only, is located almost two miles north of the freeway exit; park on the shoulder of the old highway and walk through the tunnel.

HOBSON BEACH COUNTY PARK: *Off old Pacific Coast Hwy., just S. of Sea Cliff.* This compact park offers day use and camping in a palm grove by the sea. There are picnic tables, a concession stand, restrooms with running water, hot showers, and a campground host. There are also 31 tightly packed trailer and tent campsites, ten of which have full hookups, including water, sewer, electric, and cable TV. Pets OK on maximum six-foot leash. Fee for overnight use. Call: 805-654-3951. Activities include surf fishing and surfing; the reef is good for exploring at low tide. A concrete stair provides access over the riprap revetment.

On January 10, 2005, the small community of La Conchita was hit by a massive mudslide that buried ten homes, destroyed several others, and killed ten people. This was hardly the town's first experience with landslides; in fact the flat area between the coastal bluff and the sea that provided a building pad for the town was first bulldozed by the Southern Pacific Railroad specifically to help reduce the landslide hazard to the rail line, which was constructed through the area in 1887. In 1924, the La Conchita del Mar subdivision was established. At that time, most of the land in the town was agricultural, and a ranch road was extended up the bluff. This ranch road was the locus of numerous small landslides throughout the middle of the twentieth century. Following heavy rains in the winter of 1994–1995, three segments of the ranch road collapsed. The big event, though, was a huge landslide in March of 1995 that brought 1.7 million cubic yards of the hillside crashing into town, destroying seven residences. Fortunately, there were no fatalities or serious injuries. A week later, a debris flow occurred in the canyon north of the slide, damaging an additional five homes and a banana plantation. Litigation ensued, brought about by homeowners who felt that irrigation of the citrus and avocado groves at the top of the bluff caused the slide. The court disagreed, and the homeowners were left to decide whether to rebuild or leave. Most stayed, only to witness the latest slide a decade later.

To a geologist, the bluff at La Conchita shows all the signs of being the site not only of recent landslides but also of far larger ancient landslides. The top of the bluff shows a scarp that indicates that the bulk of the bluff—perhaps 100 million tons of earth—has slipped to the sea in the past. Geologists at the University of California at Santa Barbara have suggested that ancient landslides may have been orders of magnitude larger than any recent landslide. What makes this section of the coast so dangerous? The rocks making up the bluff at La Conchita are notoriously weak, consisting of poorly consolidated shales and mudstones. They also are sheared by an active fault that cuts the bluff. Steep slopes are maintained by the high uplift rates in the region (among the highest in the world), and drainage problems saturate the slope. Whether or not La Conchita is built on a megaslide as some researchers believe, the community is clearly at high risk from landslides far larger than the 2005 mudflow.

La Conchita

Surfer off Rincon Point

Surfing has permeated the California coastal culture, and each part of the coast has its own surfing identity. Rincon in northern Ventura County has been known for surfing almost since surfing first started in California. As winter storms develop to the northwest, waves come down the coast and refract or wrap around the point at Rincon, creating almost perfect surfing breaks. South of Rincon are numerous smaller surfing areas, many of them named for local surfers or landmarks.

One such area, just south of Mussel Shoals, has had several names that reflect changes in the area over time. Stanley's was a shallow cobble reef named for the small diner that once stood close by; an artificial island was built to support oil production near Stanley's. Surfing and oil were able to coexist in this area for many years. However, oil production activities and the construction of new homes created the need for better roads, and in the 1970s, the road near Stanley's was widened. The diner was removed and the cobble reef was buried, resulting in the loss of both the surfing spot and the diner for which it was named.

Later, several oil piers were constructed seaward of Stanley's. These piers blocked some of the wave energy coming past the artificial island, causing the build-up of sand bars around the piers and establishing a new surf spot that came to be called Oil Piers. Until the late 1990s, when some of the oil facilities were decommissioned and the piers were removed, Oil Piers remained a popular surfing spot. Now the area is being considered for construction of an offshore submerged reef, to be built out of geotextile woven bags filled with sand.

The reef is being designed primarily to reduce shoreline erosion and to help sand remain on the beach for a longer time. A secondary aspect of the submerged reef is that is it being designed to improve the surf break, like the artificial surfing reefs that have been constructed along the coast in Australia. One such reef is at Gold Coast in Queensland. In the future, this location southeast of Rincon may experience yet another surfing metamorphosis—starting as a cobble reef break, changing to a sand bar break, and now, perhaps, becoming a sand bag reef break.

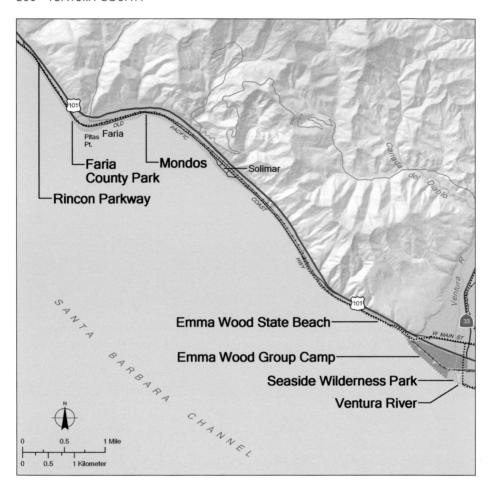

Faria County Park

Faria County Park

Rincon Parkway to Ventura River

	Sandy Beach	Rocky Shore	Trail	Visitor Center	Campground	Wildlife Viewing	Fishing or Boating	Facilities for Disabled	Food and Drink	Restrooms	Parking	Fee
Rincon Parkway	•	•	•		•		•	•		•	•	•
Faria County Park	•	•		•	•	•	•		•	•	•	•
Mondos	•									•		
Emma Wood State Beach	•				•		•			•	•	
Emma Wood Group Camp	•	•	•		•	•	•	•		•	•	•
Seaside Wilderness Park			•			•						
Ventura River			•			•						

RINCON PARKWAY: *Old Pacific Coast Hwy., between Hobson and Faria County Parks.* The roadside along the seawall provides overnight camping in designated areas for up to 127 self-contained recreational vehicles. No day-use parking allowed; no reservations. A bicycle path from Carpinteria to Ventura runs between the roadway and the campsites. Chemical toilets; fee for camping. Call: 805-654-3951. Upcoast of the RV camping area, day-use parking on the road shoulder is allowed from 7 AM to 9 PM, but watch for signs where parking is prohibited; restrictions are stringently enforced.

FARIA COUNTY PARK: *Old Pacific Coast Hwy. at Pitas Point.* This small park tucked between old Pacific Coast Hwy. and the ocean offers overnight camping and day use between 7 AM and sunset. A stairway provides access over a revetment to a long sandy beach. Park facilities include 42 tent and trailer campsites, some next to the ocean and some with full hookups. There are hot showers, a concession stand, ice machine, and horseshoe pits. Campground host; fee for camping. Good tidepooling among the rocks at low tide. Western gulls, which breed on Anacapa Island to the south, rest and feed on shore. Call: 805-654-3951. The park may be closed during heavy storms.

MONDOS: *Along old Pacific Coast Hwy., between two parts of Faria community.* Between two groups of residences in the Faria Beach community is a cove known to generations of surfers as Mondos, or Mandos, a reference to a steakhouse called Mando's Place that stood here until the early 1950s. This is a popular spot for beginners. Roadside parking is prohibited seaward of the highway but allowed on the inland side, between the hours of 7 AM and 9 PM. Use caution crossing the highway. No facilities. Additional beach access is available over the riprap revetment at two locations 500 feet north and 100 feet south of the Solimar community; daytime shoulder parking allowed.

EMMA WOOD STATE BEACH: *S. end of old Pacific Coast Hwy., N. of West Main St., Ventura.* A linear park, once part of old Hwy. One, is located between the ocean and the railroad tracks. There are 90 campsites for fully self-contained vehicles only; no tent camping; no reservations. Vehicles are limited to 30 feet in length; sites are not level. No water, electricity, or restroom facilities here; the nearest dump station facilities are available at McGrath State Beach downcoast. Park gates are open from 6 AM to 10 PM daily; fee for entry. From October 1 through March 31, camping is available Friday through Sunday only.

EMMA WOOD GROUP CAMP: *Off Main St., W. of the Ventura River bridge, Ventura.* Despite its name, this part of Emma Wood State Beach offers facilities for both day use and camping. Groups of up to 30 tent-campers and up to 50 persons in RVs can be accommodated; maximum vehicle length is 35 feet. Reservations required year-round; call 1-800-444-7275.

Two paths lead from the parking area to the shoreline. Near the park's entrance station, follow the bike path northwest 100 yards to a railroad crossing, where a short path leads to a sandy beach. A second shoreline access-way starts southwest of the main parking lot and leads under the tracks to a cobble beach. A small marsh community of tules and cat-tails is at the south end of the state beach, at the former second mouth of the Ventura River. Willets, curlews, and other shorebirds rest and feed along the shore. Anglers catch surfperch, cabezon, and California corbina; common dolphins and Pacific white-sided dolphins may be seen offshore.

SEASIDE WILDERNESS PARK: *Seaward of Hwy. 101, W. of the Ventura River, Ventura.* A small undeveloped park of Monterey cy-press and palm trees is located adjacent to the mouth of the Ventura River. Good bird-watching; double-crested cormorants roost in the trees, and brown pelicans gather at the river mouth. Access is from Emma Wood Group Camp, along the dunes seaward of the railroad tracks. A remote spot, used in the past as a transient encampment.

VENTURA RIVER: *Seaward of Hwy. 101, W. of Fairgrounds.* The Ventura River drains a 220-square-mile area including the Santa Ynez Mountains. Part of the lower river val-ley has been developed for agriculture and industry, but natural habitat remains. Ripar-ian vegetation includes arroyo willow, black willow, and tree tobacco; coastal sage scrub habitat includes salt bush, California sage-brush, and coyote brush. The Ventura River supports probably the largest of the remnant runs of steelhead trout in southern Califor-nia. The Ventura River Trail, a paved bicycle path, extends from W. Main St. at N. Olive St. in Ventura along the river a distance of six miles to sycamore-shaded Foster Park, which offers picnicking on the grass, fish-ing, and camping. Call: 805-654-3951. From Foster Park, the bicycle path continues as the Ojai Valley Trail another 11 miles to the town of Ojai.

Ventura River at Foster Park

Oil development at Summerland, Santa Barbara County, ca. 1911

Oil development in California expanded considerably during the 1890s as exploration and production moved offshore. The Summerland field in Santa Barbara County was the first offshore oil field developed in the nation, and possibly the world. Nearly 100 different operators produced from 14 piers on the beach near the town of Summerland, about seven miles east of Santa Barbara. By 1902 the operators had drilled 412 wells, but each well's output dwindled quickly and by 1903, 114 wells were idle and 100 had been deserted. Only a few wells remained active in the 1920s. Today, the oil wells on the beach at Summerland are gone; however large, modern drill rigs are visible about 12 miles offshore.

Until the late 1940s, offshore wells were drilled from piers or in shallow waters adjacent to the shoreline. In Louisiana in October 1947, the Kerr-McGee Corporation installed the first offshore platform out of sight of land, more than 10 miles from shore. This expansion of offshore development led to an intense struggle between the federal and state governments over ownership of the Outer Continental Shelf, or OCS. In 1953, the federal government enacted the Submerged Lands Act and Outer Continental Shelf Lands Act, which settled the dispute over ownership, and established the federal government's authority to lease federal submerged lands for the purpose of developing oil and gas. Shoreline states, including California, retain full rights to lease submerged lands within three miles of shore, but the area from three miles to 200 miles offshore is within the jurisdiction of the federal government. Under federal law adopted in 1972, California may participate in federal decisions regarding offshore oil and gas exploration and development.

In 1966, over local protests, oil companies acquired the first federal OCS lease in the Santa Barbara Channel. Platform Hogan was installed just beyond the state's three-mile boundary offshore Carpinteria in 1967. By 2005, 19 offshore oil platforms supported production of developed leases on the federal OCS, with a handful more located in state waters. In Ventura County, four platforms—named Grace, Gilda, Gail, and Gina—are located in federal waters offshore the city of Ventura, while at the very north edge of the county, Rincon Island produces oil and gas from state-owned tidelands.

California's annual offshore oil production peaked at 70 million barrels in 1995, and production rates have been in steady decline since then. In 2003, the California OCS produced 28 million barrels of oil—enough to supply the state of California with petroleum products for approximately two weeks. The California Oil Museum in the city of Santa Paula, 12 miles east of Ventura, presents the history and workings of the oil industry through interactive displays, videos, working models, games, photographs, restored gas station memorabilia, and an authentic turn-of-the-century cable-tool drilling rig.

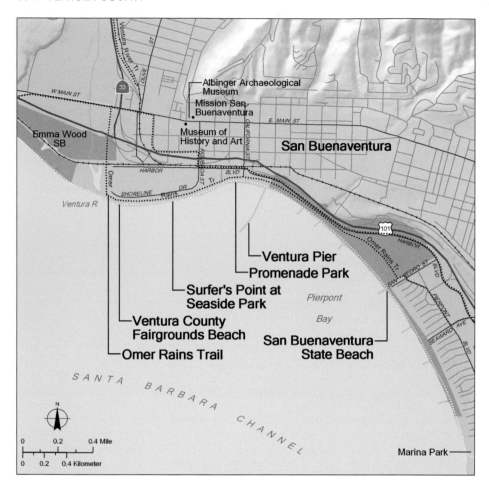

Ventura River Tr.
OLIVE ST
VENTURA ST
33
W MAIN ST

Albinger Archaeological
Museum
Mission San
Buenaventura
E. MAIN ST
Museum of
History and Art

CALIFORNIA ST

San Buenaventura

Emma Wood
SB

HARBOR
SHORELINE   DR
Omer
Rains   Tr.

FIGUEROA ST

BLVD

Ventura R

101
HARBOR
Omer Rains Tr.

SAN PEDRO ST
BLVD
PIERPONT

SEAWARD   AVE
BLVD

Ventura Pier
Promenade Park

Surfer's Point at
Seaside Park

Pierpont

Bay

Ventura County
Fairgrounds Beach

San Buenaventura
State Beach

Omer Rains Trail

S A N T A   B A R B A R A   C H A N N E L

N

0   0.2   0.4 Mile
0   0.2   0.4 Kilometer

Marina Park

Promenade Park

San Buenaventura

	Sandy Beach	Rocky Shore	Trail	Visitor Center	Campground	Wildlife Viewing	Fishing or Boating	Facilities for Disabled	Food and Drink	Restrooms	Parking	Fee
San Buenaventura	•	•	•	•	•	•	•	•	•	•	•	
Omer Rains Trail	•		•					•	•	•	•	
Ventura County Fairgrounds Beach		•	•								•	•
Surfer's Point at Seaside Park		•	•					•		•	•	
Promenade Park	•		•					•			•	
Ventura Pier						•	•	•	•	•	•	
San Buenaventura State Beach	•		•			•	•	•	•	•	•	•

SAN BUENAVENTURA: *65 mi. N.W. of Los Angeles.* The city of San Buenaventura is located on a long curve of sandy shoreline, an easy walk from the revitalized downtown that features a Spanish colonial mission church, museums, and shops. A visitor center is located at 89 S. California St., just off the Hwy. 101 freeway; call: 805-648-2075. The Ventura County Museum of History and Art, located at 100 E. Main St., has exhibits about local farming history, the Chumash indigenous people, and other early residents, along with a museum store. Open Tuesday through Sunday, 10 AM to 5 PM; call: 805-653-0323. Next to the old mission, which is an active parish church, is the Albinger Archaeological Museum at 113 E. Main St. Exhibits cover 3,500 years of Ventura history; long before the Spanish founded the original mission here in 1782, a Chumash village called Shisholop existed near the mouth of the Ventura River. Open Wednesday through Sunday; call: 805-648-5823. The modern downtown contains residences and commercial buildings in a variety of architectural styles, from 19th century Queen Anne to Spanish Colonial Revival to 20th century Moderne.

OMER RAINS TRAIL: *From Ventura River to Oxnard State Beach.* A paved path popular with bicyclists and skaters starts at a parking lot on W. Main St., east of the Ventura River bridge. The trail follows the river south to the sea and then turns east along the shoreline, past the Ventura Pier, through San Buenaventura State Beach, and beyond, to Ventura Harbor and Oxnard. From the terminus of the Omer Rains trail, a bicycle path also leads west across the Ventura River bridge, through Emma Wood State Beach, and along old Pacific Coast Hwy. to the Santa Barbara County line, a distance of 12 miles. Near Mussel Shoals, where there is no frontage road, the bike lanes run along the shoulder of Hwy. 101.

VENTURA COUNTY FAIRGROUNDS BEACH: *Between the Ventura River and Surfer's Point at Seaside Park, Ventura.* The rocky cobble beach is accessible from Surfer's Point downcoast or along the Omer Rains bicycle trail from W. Main St. The beach is very popular with surfers and kiteboarders. Fee parking is located next to the beach; take Figueroa St. south to its end, then head west. The Ventura County Fair runs annually for approximately one week at the beginning of October; other events are held frequently. Call: 805-648-3376. Beach erosion has caused portions of the shoreline bicycle path to be moved inland; continuing erosion may cause future changes in the location of parking lots and other facilities.

SURFER'S POINT AT SEASIDE PARK: *Foot of Figueroa St., Ventura.* A highly popular surfing spot is located east of the mouth of the Ventura River. The beach is covered with cobbles and driftwood, and it is backed by the paved Omer Rains Trail. Facilities include parking, picnic tables, outdoor showers, and restrooms, set among lawns and palm trees.

PROMENADE PARK: *From foot of Figueroa St. E. to San Buenaventura State Beach, Ventura.* A long, narrow park with a concrete walkway and bike path borders the edge of the beach. Facilities include benches, tables, playground equipment, volleyball standards, and wide ramps to the beach. Park at Surfer's Point, in the parking structure at the foot of California St., or in the lot east of the Ventura Pier.

VENTURA PIER: *E. of the foot of California St., Ventura.* The 1,700-foot-long pier is located within San Buenaventura State Beach. There is a restaurant and a grill on the pier. Bike rentals are available along the bike path, to the east of the pier. The wooden pier is wheelchair accessible, although the rough planks are bumpy. The pier is open from 7 AM to sunset.

SAN BUENAVENTURA STATE BEACH: *From the Ventura Pier S. to Marina Park, Ventura.* A wide, sandy beach extends east of downtown Ventura. The ocean waters are noted for good swimming; lifeguards are on duty daily during the summer and on weekends in spring and fall. Facilities include picnic areas set among expansive lawns, outdoor showers, modern restroom facilities, and large parking lots. Seasonally available are bicycle rentals near the main parking area and yellow beach umbrella rentals near the foot of San Jon Rd. From the parking area, marked paths channel visitors to the beach in order to protect sensitive habitat; look for killdeer feeding among the dunes. Additional access to San Buenaventura State Beach is available at the end of Seaward Ave., where there is a small, time-limited parking lot, restrooms, and a beach shower, and at 24 residential street ends between San Pedro St. and Marina Park. Call: 805-968-1033.

According to anthropologist Brian Fagan, five hundred years ago Chumash society was among the most elaborate of all hunter-gatherer societies in North America. The Chumash maritime culture embraced approximately 15,000 people in temporary and permanent settlements and relied on extensive trading contacts along the mainland coast between the modern communities of Malibu and Paso Robles and the northern Channel Islands. Chumash Indians are especially known for their baskets, woven tightly enough to hold water, their shell money, and their seafaring plank canoes called *tomols*.

The Chumash exported their excellent baskets to other tribes even in pre-European times. After the Spanish entrada, explorers and settlers sent baskets as souvenirs to family and friends in all parts of the world. Today there are about 400 Chumash baskets held worldwide in museums and private collections. Chumash basket designs are easy to recognize: most have a "principal band," a sort of border about an inch wide, below the rim. Below that, designs might include vertical bars, horizontal bands, zigzags, stepped lines, or an all-over network pattern. The Ventura County Museum of History and Art keeps a collection of Chumash baskets and other objects, while the Santa Barbara Museum of Natural History has the largest collection of Chumash baskets in the world.

Shell beads developed as currency about 850 years ago. At that time nearly all shell beads were made in villages on the northern Channel Islands. Shell bead money, or *'Anchum*, was usually made from small disks shaped from the shell of a marine snail. The value of the money depended on the labor invested to make it and the rarity of the shell that was used. The disk beads made from the callus, the thick part of the shell near its opening, were worth twice as much as the disk beads made from the wall of the shell, because many more beads could be made from the wall. The value of a strand of beads was measured according to its length—how many times it would wrap around a person's hand.

The Chumash traveled routinely between the islands and mainland, using *tomols* as fishing vessels and to trade acorns and food from the mainland for shell beads crafted on the islands. To make the *tomols*, the Chumash used redwood logs that floated down the coast and washed up on the beaches. Planks made from the logs were lashed together with cord made from Indian hemp and sealed with a waxy mixture of natural asphalt and pine pitch. *Tomols* were decorated with paint and shell inlay, and occasionally with powdered shell thrown against the wet finish before it dried.

The last Chumash *tomols* used for fishing were made in about 1850. In 1913, an elderly Chumash man made a *tomol* for anthropologist John Harrington, to show how they were built. That boat is now on exhibit at the Santa Barbara Museum of Natural History. In the past twenty years several *tomols* have been made using Harrington's notes to guide construction. In 2001, members of the Chumash tribe crossed the Santa Barbara Channel in a *tomol* for the first time in over 125 years. Even with modern safety equipment, the twenty-plus-mile journey was a formidable and grueling venture. Two other crossings have been made since, in 2004 and 2005, and the tribe hopes the crossing will become an annual event.

Ceremonial reenactment of Chumash *tomol* voyage across Santa Barbara Channel

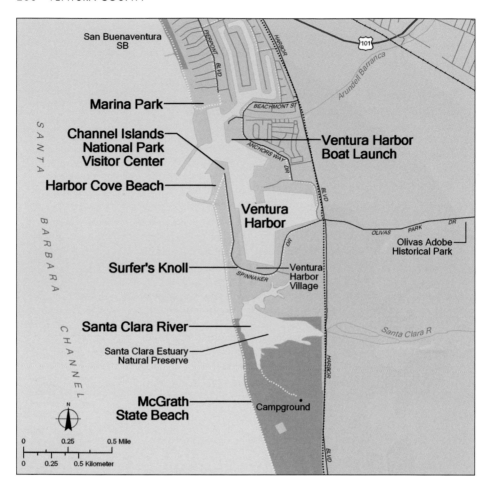

San Buenaventura
SB

SANTA

Marina Park

Channel Islands
National Park
Visitor Center

Harbor Cove Beach

BARBARA

Ventura Harbor
Boat Launch

Ventura
Harbor

Surfer's Knoll

CHANNEL

Santa Clara River

Santa Clara Estuary
Natural Preserve

McGrath
State Beach

Ventura
Harbor
Village

Campground

Olivas Adobe
Historical Park

Santa Clara R

PIERPONT BLVD

HARBOR

BEACHMONT ST

ANCHORS WAY

DR

BLVD

SPINNAKER

DR

OLIVAS   PARK   DR

HARBOR

BLVD

Arundell Barranca

101

N

0   0.25   0.5 Mile

0   0.25   0.5 Kilometer

Ventura Harbor

Ventura Harbor Area

	Sandy Beach	Rocky Shore	Trail	Visitor Center	Campground	Wildlife Viewing	Fishing or Boating	Facilities for Disabled	Food and Drink	Restrooms	Parking	Fee
Marina Park	•					•	•			•	•	
Ventura Harbor Boat Launch							•	•	•	•	•	
Ventura Harbor	•					•	•	•	•	•	•	•
Channel Islands National Park Visitor Center				•		•		•		•	•	
Harbor Cove Beach	•						•			•	•	
Surfer's Knoll	•	•					•			•	•	
Santa Clara River		•				•						
McGrath State Beach	•		•	•	•	•	•	•		•	•	•

MARINA PARK: *S. end of Pierpont Blvd., Ventura.* This city park offers access to both the harbor, sheltered by breakwaters, and an ocean beach. Facilities include picnic sites, an unusual ship play structure, sand volleyball court, and restrooms. On the harbor side of the park is a dock for fishing and hand-launching of boats. Sailing lessons are offered with advance reservations. For information, call: 805-652-4550.

VENTURA HARBOR BOAT LAUNCH: *Anchors Way Dr., Ventura Harbor.* A six-lane boat ramp is open 24 hours a day; no launching fee, but vehicles must obtain a parking permit from the dispenser at the restroom building; first hour free with permit. Next to the boat ramp are picnic tables overlooking the harbor and a store offering bait and marine fuel. Dump station for boats available.

VENTURA HARBOR: *W. of Harbor Blvd. at Anchors Way Dr., Ventura.* Extensive inland harbor with marinas, boat charters, fuel dock, boat storage and repair, small sailboat rentals, and other boating facilities. The Ventura Harbor Village commercial center on Spinnaker Dr. offers restaurants, shops, and waterfront walkways. Call: 805-644-0169.

CHANNEL ISLANDS NATIONAL PARK VISITOR CENTER: *End of Spinnaker Dr., Ventura Harbor.* The main site, on the mainland, for information about the Channel Islands National Park, which includes Santa Barbara Island and the four Northern Channel Islands: San Miguel, Santa Rosa, Santa Cruz, and Anacapa. The visitor center offers interpretive exhibits and programs, a tidepool display, native plant garden, bookstore, and a picnic area overlooking Ventura Harbor. Island Packers, located in the harbor near the visitor center, provides transportation to the islands of the national park; for information, call: 805-642-1393.

During the summer months, visitors to Ventura Harbor can view the underwater video program that takes place on Anacapa

Marina Park

Santa Clara Estuary Natural Preserve

Island. On Tuesdays and Thursdays at 2 PM, a diver equipped with a video camera and microphone explores the kelp forest near the island, while visitors on shore view the surroundings. Open daily, except Thanksgiving and Christmas. There are also Channel Islands National Park visitor centers on the islands of Anacapa and Santa Barbara.

HARBOR COVE BEACH: *Spinnaker Dr., Ventura Harbor.* Jetties shelter this swimming beach. Summer lifeguard service is provided by the city of San Buenaventura. Restrooms are located near the parking lot. Call: 805-652-4550. South of the parking lot, outside the breakwater, is a wide sandy beach where rip currents can be hazardous; swimming is not advised.

SURFER'S KNOLL: *Spinnaker Dr., Ventura Harbor.* A parking area with restrooms and beach shower is located west of Spinnaker Dr., opposite the Ventura Harbor Village. Wide, sandy Peninsula Beach extends south to the mouth of the Santa Clara River. East of the beach are several trails leading around the city's water reclamation ponds; a good birding spot.

SANTA CLARA RIVER: *The mouth of the Santa Clara River is a natural preserve within McGrath State Beach.* The riverbanks support a riparian habitat of sandbar willow and black cottonwood; the invasive, non-native giant reed called arundo is widespread. A coastal sage scrub community of saltbush, coyote brush, and California sagebrush grows along the river, and freshwater marsh plants such as tule, cattail, and bulrush grow in the river channel.

Wildlife along the river includes painted lady butterflies and anise swallowtail butterflies; the tidewater goby is found in the river, and remnants of the Santa Clara River's once abundant steelhead population spawn upstream.

The State Coastal Conservancy, along with the Friends of the Santa Clara River and the Nature Conservancy, is working to create the Santa Clara River Parkway. This project includes restoration of natural river processes and aquatic and riparian habitat, designed to prevent flooding and to improve conditions for steelhead and other native species, including birds that migrate along the river corridor. Flooding has been exacerbated by narrowing of the river's natural channel, resulting from agricultural and urban development; implementation of the parkway will include widening of the channel through removal of levees. The parkway will also include a continuous public trail system, reaching some 24 miles inland to the confluence of Sespe Creek.

MCGRATH STATE BEACH: *Along Harbor Blvd., 1.2 mi. S. of Spinnaker Dr.* The state park includes a campground and two miles of very wide, sandy beach. Surfing and fishing are popular, but swimmers should use caution, due to strong currents. Lifeguards on duty during the summer. Call ahead to reserve a beach wheelchair: 805-968-1033.

A nature trail, starting north of the entrance kiosk, leads into the Santa Clara Estuary Natural Preserve, a favored birding area. The sand dunes in back of the beach provide a nesting area for the California least tern. Look for sand verbena blooming in spring and summer.

There are a total of 174 family, group, and hike or bike campsites, well spaced among the trees. Sites have picnic tables and firepits; some are wheelchair accessible. For RVs, maximum length is 34 feet, and for trailers, 30 feet. RV dump station is available. Park gates are open 8 AM to 10 PM year-round; no access after hours. Day use and camping fees apply. For camping reservations, call: 1-800-444-7275. For park information, call: 805-654-4744.

McGrath Lake wildlife area is at the south end of the park, reached by walking along the beach dunes west of the oil company facilities. Because the lake's water level is maintained artificially, the shoreline has a well-established marsh community of willow, bulrush, and tule. Kelp flies, tiger beetles, Pismo clams, and littleneck clams can be found on the beach. Surfperch and bass inhabit the offshore waters; grunion spawn on the beach in spring and summer.

The Santa Clara River is one of the few in southern California without major dams and alterations. The river originates in the San Gabriel Mountains at elevations of over 8,800 feet, nearly 100 miles from the sea. Flows are highly variable, ranging from no surface flow at all during some summer months to wintertime floods that have destroyed bridges and facilities at Ventura Harbor. On March 12, 1928, the 180-foot-high St. Francis Dam, constructed by the Los Angeles Department of Water and Power on a tributary of the Santa Clara River near Saugus, failed abruptly. The collapse of the concrete dam set loose a wall of water that washed 50 miles to the sea. Over 500 persons lost their lives, and roads, bridges, livestock, and orchards were swept away.

If you cross the Santa Clara River at almost any time other than the few days after a major storm, you will see a river that appears to be carrying more sand than water. A tiny stream channel meanders among numerous sand bars forming a wide, braided channel plain. In fact, the Santa Clara River is by far the greatest supplier of sediment to the coast in southern California. The Santa Clara River discharges an average of nearly four million tons of sediment per year, although this number varies from near zero in dry years to as much as 46 million tons in wet years. The reason that this river is so prodigious is simple: it drains an area of the Transverse Ranges that is actively undergoing uplift, maintaining steep slopes and providing a ready source of erodible sediments. The river is relatively pristine, without sediment-trapping dams or concrete channelization. Furthermore, the watershed of the river is dominated by chaparral, which is highly subject to wildfires. When the chaparral burns, mudslides and erosion increase, and an especially big pulse of sediment is delivered to the ocean. As these sediments move along the coast, they help build up the beaches.

Ocean Currents and Sand Transport

LIKE GRAY WHALES offshore or monarch butterflies that migrate overland, the sand that makes up California beaches is usually in transit. Chances are that sand you see at the beach today was at a different beach last year and will be at yet another location next year. Years ago, Douglas Inman, a professor at the Scripps Institution of Oceanography, used the term "river of sand" to describe this movement of coastal sand. It is still an apt description.

Beach sand comes from various sources. Much of the sand along the Central Coast of California is brought to the shore by rivers. Every year, high flow events on the Carmel, Arroyo Grande, Santa Maria, Santa Ynez, Ventura, and Santa Clara Rivers together carry about two million cubic yards of sediment to the coast. As the velocity of a river's flow decreases at its junction with the ocean, the river is no longer able to carry enormous volumes of sand. Sand particles drop out of suspension, forming a river delta at the mouth of the river. Ocean waves sort and stir up these sand particles and carry many of them to nearby beaches. River deltas both provide sand for beaches and block some of the more erosive waves from reaching the shoreline; the result is a bulge in the beach often seen near a river mouth.

Sands on a dry beach can be moved by either wind or waves. Movement by wind, known as aeolian transport, is the main method for dune development. High winds suspend dry sand in the air and blast it inland, where it can enlarge an existing dune system. Winds can also roll or propel the sand along the beach surface in a sort of skipping motion, called saltation. Morro, Pismo, Nipomo, Jalama, and Guadalupe Dunes are all formed and maintained by the sands carried inland by persistent onshore winds. Winds carry about 200,000 cubic yards, or 270,000 tons, of sand annually onto the 35 miles of dunes between Pismo Beach and Point Arguello.

Beach sand that is not carried inland by wind will likely be carried back offshore again by regular wave action. In the surf zone, the sand will be carried up or down coast and again will be washed onto the beach some distance away from where it had been previously. Along some sections of the coast, sand moves up and down the coast regularly and can pass by the same spot two or three times as it journeys ultimately downcoast. In San Luis Obispo, Santa Barbara, and Ventura Counties, sand transport is mainly unidirectional, to the south and east. From the Santa Maria and Santa Ynez River mouths, sand travels around Point Conception and along the coast. Along the way, the river of sand carries about 250,000 cubic yards of sand past Santa Barbara each year. At Ventura Harbor, the river of sand grows to some 600,000 cubic yards of sand annually.

It may take decades for sand to move from an inland area through the river system, onto the beach, and then down the coast. Many coastal rivers have been dammed to create reservoirs for municipal and irrigation water, produce hydroelectric power, and reduce risks from flooding. Dams also hold back most of the sediment that rivers carry. When dams and flood control structures reduce a river's velocity, the amount of sand that the river can carry is also reduced. In Santa Barbara and Ventura Counties, dams and other diversion structures have reduced the amount of sediment reaching the coast by over 40 percent, 74 million cubic yards of sand is trapped by dams in the two counties.

The Matilija Dam, located on a tributary of the Ventura River approximately 16 miles from the coast, is a significant sediment trap. The dam was completed in 1947, primarily to store water for agriculture, and also for limited flood control. However, since its construction, the dam has trapped over six million cubic yards of sediment, reducing the amount of sand delivered to beaches and blocking passage of anadromous fish that historically had gone upstream to spawn. Due to the volume of sediment caught behind it, the dam long ago ceased to carry out its intended functions. The U.S. Army Corps of Engineers and the Ventura County Watershed Protection District have studied this dam and the creek habitat and have recommended that the dam be removed. Removal of the dam, which could start as early as 2009, will allow fish to return to their historic spawning and rearing habitats and will provide a more natural system in which sediment passes through the river to the beaches.

Dams are not the only barrier to sediment transport. Harbors, breakwaters, jetties, and groins can hold sand and change its movement and deposition. The breakwater at Santa Barbara Harbor has created a dramatic change to sand movement along the coast. In 1928, the breakwater was started off Point Castillo to provide safe anchorage in its lee. Leadbetter Beach upcoast and west of the breakwater, began to widen almost as soon as the breakwater was connected to the shoreline. Within four years, the western edge of the breakwater was completely buried in sand, the shoreline was realigned, and sand began traveling around the long portion of breakwater and depositing in the harbor. This reduced the utility of the harbor for boating and navigation.

At the same time that the beaches west of Santa Barbara's breakwater were growing and the harbor was filling in with sand, the beaches to the east, or downcoast, were eroding. The shore changed from sandy beach to narrow bands of boulders and cobbles, which also gradually moved downcoast. The strength of this transport was demonstrated clearly in Summerland, where in 1931 a small boiler broke through the wharf and fell onto the beach. Loose rock was placed in the boiler in the vain hope of keeping it in place, yet by 1938 the heavy object had moved 900 feet to the east. Santa Barbara Harbor has had regular dredging since 1935, and the downcoast sand transport has been reestablished mechanically; however, some sand deficit is still apparent on downcoast beaches.

Most coastal sand ends its coastal journey at one of the submarine canyons located offshore. Sand from the Carmel River forms small pocket beaches contained within two points or headlands. When these sands leave the beach, they are carried into the Carmel Submarine Canyon where waters are too deep for waves to bring them back onto land. Sands traveling east through Santa Barbara and Ventura Counties are diverted offshore Port Hueneme into two submarine canyons. Up to one million cubic yards of sand are trapped each year by Hueneme and Mugu Canyons. Once in the canyons, sand continues to travel across the continental shelf and onto the abyssal plain, and over millions of years, it may recycle as terrestrial material.

Point Mugu Beach

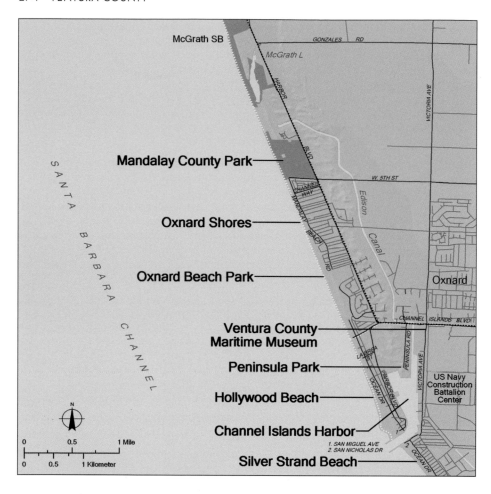

McGrath SB
McGrath L
GONZALES   RD
HARBOR
BLVD
VICTORIA AVE

Mandalay County Park
W. 5TH ST
CHANNEL WAY
Oxnard Shores
MANDALAY BEACH
Edison
Canal

Oxnard Beach Park
RD
Oxnard

S A N T A
B A R B A R A
C H A N N E L

Ventura County
Maritime Museum
CHANNEL ISLANDS BLVD
PENINSULA RD

Peninsula Park
LA BREA ST
OCEAN DR
HARBOR BLVD
VICTORIA AVE
US Navy
Construction
Battalion
Center

Hollywood Beach

N

Channel Islands Harbor
0   0.5   1 Mile
0   0.5   1 Kilometer
1. SAN MIGUEL AVE
2. SAN NICHOLAS DR

Silver Strand Beach
OCEAN DR

Peninsula Park

Oxnard Area

	Sandy Beach	Rocky Shore	Trail	Visitor Center	Campground	Wildlife Viewing	Fishing or Boating	Facilities for Disabled	Food and Drink	Restrooms	Parking	Fee
Mandalay County Park	•					•	•					
Oxnard Shores	•											
Oxnard Beach Park	•		•					•		•	•	•
Ventura County Maritime Museum				•				•		•	•	
Peninsula Park							•			•	•	
Hollywood Beach	•						•	•		•	•	
Channel Islands Harbor			•			•	•	•	•	•	•	•
Silver Strand Beach	•					•	•		•	•	•	

MANDALAY COUNTY PARK: *End of W. 5th St., W. of Harbor Blvd., Oxnard.* An undeveloped stretch of beach and dunes is located on the north side of W. 5th St.; roadside parking. The beach is used for fishing, surfing, and strolling. There is lots of driftwood, and western gulls, terns, and shorebirds may be seen here. A restoration program is underway in the dunes, which are vegetated with beach primrose, sea rocket, and sea fig. The power plant facility to the north is private property; do not trespass.

OXNARD SHORES: *Along Mandalay Beach Rd., S. of W. 5th St., Oxnard.* This residential area, located west of Harbor Blvd., has two public beach areas: at the north end on Mandalay Beach Rd. between W. 5th St. and Channel Way, and near the south end at Neptune Square. Public beach access is also available between homes at several signed locations. Street parking; no facilities.

OXNARD BEACH PARK: *W. of Harbor Blvd., 1 mi. S. of W. 5th St., Oxnard.* This nicely maintained day-use park includes palm-studded lawns, a dune field, and a sandy beach. There are individual and group picnic areas, including a covered pavilion, barbecue grills, play equipment, and restrooms. A wheelchair-accessible paved loop path leads through the dunes to an additional picnic area overlooking the sea.

VENTURA COUNTY MARITIME MUSEUM: *2731 S. Victoria Ave. at Channel Islands Blvd., Oxnard.* The museum has a sterling collection of paintings of maritime subjects, including battle scenes and works by 17th century Dutch and Flemish artists. The collection of historic ship models includes elegantly detailed vessels made during the Napoleonic Wars in Europe. Open daily from 11 AM to 5 PM; donations appreciated. Call: 805-984-6260.

PENINSULA PARK: *Peninsula Rd., .5 mi. S. of Channel Islands Blvd.* This harbor-front park has a sandy playground area, boat dock, grassy picnic area with barbecue grills, and two tennis courts. Parking and restrooms are adjacent.

HOLLYWOOD BEACH: *Along Ocean Dr., between Channel Islands Blvd. and San Miguel Ave., Oxnard.* A very wide, white sand beach lies seaward of the homes, north of the ocean entrance to Channel Islands Harbor. Beach access is via the street ends along Ocean Dr. Restrooms with outdoor showers and a small parking lot are at the corner of Ocean Dr. and La Brea St. Volleyball standards and nets are available. Lifeguards on duty during the summer only.

CHANNEL ISLANDS HARBOR: *W. of Victoria Ave., S. of Channel Islands Blvd., Oxnard.* Channel Islands Harbor, once dune fields and wetlands, was excavated in 1960 by the U.S. Army Corps of Engineers, which also constructed entrance jetties and a detached breakwater across from the harbor entrance.

Silver Strand Beach

Recreational outfitters near Channel Islands Harbor include:
Channel Islands Kayak Center, 805-984-5995.
Hopper Boat Rentals, 805-382-1100.

Surf shops near the beach include:
Momentum Surf Company, 805-985-4929.
Anacapa Surf N Sport, 805-488-2702.

Waterworks, 805-984-9283.

Bicycle rentals near Oxnard Beach Park:
Freetime, 805-984-1994.
Wheel Fun Rentals, 805-650-7770.

Whale-watching and fishing trips:
Island Packers, 805-382-1779 or 805-642-7688.
Channel Islands Sportsfishing, 805-382-1612.

The harbor contains over 2,000 boat slips, commercial fishing facilities, boat ramps and hoists, sport fishing and whale-watching services, and bicycle and pedestrian paths along the waterfront.

SILVER STRAND BEACH: *Along Ocean Dr., S. of San Nicholas Ave., Oxnard.* A wide sandy beach lies seaward of the residential area along Ocean Dr., between the ocean entrances to Channel Islands Harbor and the Port of Hueneme. The main access is at the end of San Nicholas Ave., where there are restrooms, beach volleyball standards, and parking. Lifeguards on duty summers only. A wheelchair-accessible ramp overlooks the harbor entrance. Pets must be leashed. A second parking lot is at the south end of Silver Strand Beach, at Sawtelle Ave.; the gate closes at sundown. The jetty extending seaward of the beach is the rock-filled shipwreck of the ocean liner S.S. *La Jenelle*, now used for fishing; the jetty may be hazardous during heavy surf.

Tens of thousands of years ago, during the last ice age, sea level was as much as 400 feet lower than at present, and the four northern Channel Islands were connected into one land mass known as Santarosae, its eastern tip separated by only a few miles from what is now Ventura County. Santarosae was colonized by Columbian mammoths that swam across this channel in search of new foraging grounds. The Columbian mammoth, *Mammuthus columbi*, once ranged across the present United States and as far south as Nicaragua. Standing over 14 feet tall, they were much larger than modern elephants, and subsisted mostly on a diet of grasses and sedges.

The remains of numerous individual Columbian mammoths have been recovered from the northern Channel Islands, mostly from the relatively flat marine terraces rimming the islands. Far more numerous, however, are the remains of a race of pygmy mammoths that are found on the islands but are unknown on the mainland. Standing four to eight feet tall, they were only half the size of the contemporaneous Columbian mammoth. As sea level rose, the Columbian mammoths were isolated from the mainland, and they evolved into the pygmy form, *Mammuthus exilis*. The steep island interiors and limited habitat on the islands favored the smaller mammoths, which were able to negotiate slopes as much as ten degrees steeper than the larger mammoths could handle. Perhaps this capability opened more territory to them, allowing them to thrive on the islands even as sea level rose. The remains of pygmy mammoths have been found on San Miguel, Santa Rosa, and Santa Cruz Islands.

The most dramatic mammoth discovery was made in 1994 when a nearly complete, articulated pygmy mammoth skeleton was uncovered on Santa Rosa Island. This skeleton, along with recreation of how it appeared *in situ*, and the remains of other pygmy mammoths can be seen at the Santa Barbara Museum of Natural History. Radiocarbon dating shows that this animal lived about 13,000 years ago. Apparently, pygmy mammoths coexisted with the ancestral Chumash at about this time, but became extinct shortly thereafter. Whether hunting by the Chumash, competition among species, climate change, or some other cause led to their extinction remains unknown.

Pygmy mammoth skeleton

Beaches

I N ADDITION to carrying sand, ocean waves bring floating debris on to beaches and leave it on the sand, where it serves as a general indicator of the inland extent of wave run-up. The washed-up organic material, including wood, seaweeds, and seagrasses, is collectively called wrack, and the strip of wave-cast material is called the wrack line. This wrack line is one of the most productive ecological zones on the beach. It is also a key link between marine and terrestrial ecosystems. Most of the wrack comes from the ocean, although heavy rains also carry terrestrial vegetation to the beach. The wrack provides food and shelter for beach hoppers, flies, and isopods that form the basis of a food chain that feeds beach beetles, small invertebrates, and shorebirds, such as the black-bellied plover, western snowy plover, and long-billed curlew. The highest wrack line is also the zone where California grunion spawn during spring and summer, and their eggs incubate in the sand before hatching and washing into the ocean during the next full-moon or new-moon high tide.

Scientists have only recently recognized the full ecological importance of beach wrack. Researchers at the University of California at Santa Barbara have found significant differences among beaches with and without wrack, suggesting that the wrack food chain provides a smorgasbord of prey for a multitude of shorebirds. The wrack also holds sand that forms small seasonal dunes, which may sprout vegetation and serve as wave breaks.

Unlike other species, humans have not appreciated the value and richness of beach wrack. Along more than one hundred miles of recreational beaches in California, beach managers groom the sand with large mechanized equipment, raking up trash, debris, and wrack in an effort to make the beach more visually appealing, to cut down on small insects and beach hoppers that live in the wrack, and to reduce the tangy aroma that can develop as the wrack ages. Mechanical removal of the wrack also removes the foundation of the food web and an important building block for the beach ecosystem. Efforts are under way at a few beaches to apply hand grooming or zonal grooming to remove most of the trash and litter that collect on the beach, while leaving behind the natural wrack. You can visit groomed and ungroomed beaches to see these differences for yourself. Beaches in Ventura County that are typically groomed include Silver Strand Beach, Mandalay County Park, and most of San Buenaventura State Beach. In Santa Barbara County, grooming is practiced at Carpinteria City Beach and East Beach, West Beach, and Leadbetter Beach in Santa Barbara. Examples of ungroomed beaches include Surfer's Knoll and the eastern end of Ormond Beach, both in Ventura County; Arroyo Burro Beach County Park, Refugio State Beach, Gaviota State Park, and Jalama Beach County Park, all in Santa Barbara County. Beaches in San Luis Obispo County and the majority of beaches in Monterey County are left ungroomed.

Long-billed curlew feeding in beach wrack

The sandy beach environment is not an easy place for organisms to live. Animals have to deal with crashing waves, changing tides, a beach that changes seasonally, and marine and terrestrial predators. Some of the animals that live in this environment are buried in the sand. This is where the **spiny mole crab** (*Blepharipoda occidentalis*) can be found. It generally occurs in the subtidal part of the beach. The spiny mole crab can reach about two and a half inches in length and can burrow very rapidly. As the spiny mole crab grows, the exoskeleton, or shell, is molted periodically. While you are walking along the beach, look for the molts, or disposed exoskeletons, from the spiny mole crab. The antennae you see poking out of the sand in the swash zone are of another burrowing species, the sand crab (*Emerita analoga*).

Spiny mole crab

Pismo clams (*Tivella stultorum*) are bivalve mollusks that inhabit burrows on flat beaches in the intertidal zone and offshore to a depth of 80 feet. Pismo clams are usually found buried about two to six inches deep. They take in water through a siphon, and the water is passed over gills where food particles are removed and then consumed by the clam. The Pismo clam feeds on detritus (pieces of disintegrating plants and animals), phytoplankton (such as diatoms), bacteria, and zooplankton. Until the early 1970s, the Pismo clam was one of the most abundant clam species in the lower intertidal zone along the beaches of the central California coast. However, two factors combined to result in a drastic decrease in numbers: over-harvesting, and sea otter predation. If you find a clamshell washed up on the beach, look to see if the shell is broken diagonally; if so, it might have served as lunch for a hungry sea otter.

Pismo clam

Among the most widespread of all shorebirds, **sanderlings** (*Calidris alba*) turn up on beaches all over the world. As a wave comes roaring in, these cute little birds run up just ahead of the breaker, then sprint after the retreating wave to feed on the tiny crustaceans and mollusks left exposed. This is the only sandpiper that lacks a hind toe, allowing it to be a strong runner. Watch for the sanderlings racing with the waves next time you go to the beach.

Sanderling

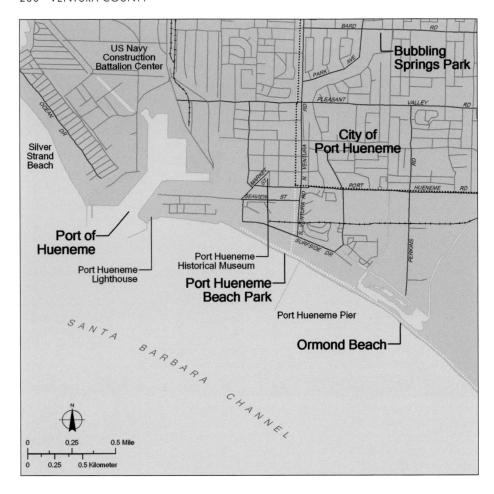

Wetlands at Ormond Beach

Port Hueneme Area

	Sandy Beach	Rocky Shore	Trail	Visitor Center	Campground	Wildlife Viewing	Fishing or Boating	Facilities for Disabled	Food and Drink	Restrooms	Parking	Fee
Port of Hueneme							•					
City of Port Hueneme	•		•	•			•	•	•	•	•	
Port Hueneme Beach Park	•		•				•	•	•	•	•	•
Ormond Beach	•		•			•					•	
Bubbling Springs Park			•					•		•	•	

PORT OF HUENEME: *W. end of Hueneme Rd., Port Hueneme.* The Port of Hueneme, the only deepwater harbor between San Francisco and San Pedro, was dredged from 300 acres of land and opened in 1940. During World War II, the U.S. Navy took over the port, and 200,000 military personnel shipped out of Port Hueneme. After the war, the Oxnard Harbor District gradually reacquired part of the harbor, and the facility became a major support base for offshore oil facilities in the Santa Barbara Channel. Goods shipped through the harbor include automobiles, fresh produce, and forest products; the Port of Hueneme leads all other U.S. seaports in the export of citrus fruit. Port Hueneme Sportfishing is located at the port; call: 805-488-2212. For information about the port, call: 805-488-3677.

Hueneme, pronounced "why-nee-mee," is derived from a Chumash word meaning "half-way" or "resting place." The place to which the name is applied is midway between Point Mugu and the mouth of the Santa Clara River; it is also the closest place on the mainland to the Channel Islands (the tip of Anacapa Island lies 11 miles away). East of the harbor is the U.S. Navy Construction Battalion ("Seabee") Center. Within the navy base is the Seabee Museum, which contains war memorabilia and other artifacts; call: 805-982-5165.

CITY OF PORT HUENEME: *N. and E. of Port Hueneme Harbor.* The Port Hueneme Historical Museum, at 220 N. Market St., is open weekdays from 9 AM to 3 PM, and closed between 12 noon and 1 PM; call: 805-488-2023.

Port of Hueneme

The Port Hueneme Lighthouse, built in 1874 and rebuilt in 1941 in Moderne architectural style, is open on the third Saturday of the month, from 10:30 AM to 3:00 PM. The lighthouse, which is operational, is located on property of the Port of Hueneme near the west end of Hueneme Rd.; for information, call: 805-488-2023.

PORT HUENEME BEACH PARK: *S. end of Ventura Rd., Port Hueneme.* A broad, sandy beach extends seaward of Surfside Dr. There are protected picnic sites, with barbecue grills, set in a grassy area and along the beach. A paved pedestrian and bike path runs along the edge of the beach. Fee for parking; some parking spots are open 24 hours. For park information, call: 805-986-6555. At the north end of the park is a memorial to those lost at sea on Alaska Airlines flight #261 in 2000.

The 1,240-foot-long Port Hueneme Pier is lighted and open 24 hours. Once T-shaped, the pier lost its outer end in storms in 1995 and 1998; reconstruction work has been funded by the State Coastal Conservancy in conjunction with the city of Port Hueneme and the Wildlife Conservation Board. The pier is popular with anglers; there are cutting tables with sinks and a bait shop with fishing supplies. Spider crabs are caught here. The pier is the site of the annual Hueneme Beach Festival.

ORMOND BEACH: *Foot of Perkins Rd., S. of Port Hueneme.* As a recreational resource and wildlife habitat, this area is a work in progress. Accessible now from the parking area at the end of Perkins Rd. are low dunes and a wide sandy beach; more facilities are on the way. Ormond Beach can be reached also by walking south along the shore from Port Hueneme Beach Park, a half-mile away. The State Coastal Conservancy is pursuing a major restoration project of wetlands formerly used for industrial purposes; a re-created coastal lagoon is planned.

BUBBLING SPRINGS PARK: *Bard Rd. .5 mi. E. of N. Ventura Rd., Port Hueneme.* A grassy inland park with picnic tables, firepits, and a playground area. A bike path runs one and a half miles along a landscaped drainage channel to Port Hueneme Beach Park. Additional parking is off Park Ave. east of N. Ventura Rd.

Bubbling Springs Park

The state of California owns all tidelands, submerged lands, and the beds of inland navigable waters, holding them for "public trust" uses that include fishing, navigation, commerce, nature preserves, swimming, boating, and walking. Tidelands consist of the area on a beach or rocky coastline that is between the mean high tide line and the mean low tide line. The California Constitution guarantees the public's access to tidelands. The state or other trustee of tidelands, however, may choose one public trust use over another and generally impose reasonable restrictions on the time, place, and manner of the use of tidelands.

The boundary between public tidelands and upland property is known as the "ordinary high water mark." (California Civil Code, §§ 670, 830.) In California, on a shoreline in a natural condition, the ordinary high water mark is synonymous with the "mean high tide line," the location at which the plane of mean high water intersects with the shore. The elevation of mean high water is relatively constant; it is determined by calculating the average height of all the twice-daily high tides that occur over a 19-year period. The mean high tide line is referred to as "ambulatory," because it moves in response to changes in sand supply and beach slope. On a rocky shoreline, the location of the mean high tide line rarely changes. On a sandy beach, the mean high tide line may shift one hundred or more feet seaward as sand accumulates during the summer months, and then a similar amount landward as winter storms wash away the sand. The services of a professional land surveyor are necessary to determine the precise location of the mean high tide line at any particular time. On a sandy beach, a survey has limited value because it only establishes where the mean high tide line was located at one time.

Are there any clues for the public about the mean high tide line's location? Sometimes property owners post signs purporting to identify the mean high tide line, but those signs are inexact because of the shifting nature of the line. Informally, the wet sand is sometimes assumed to be public property; however, that reference is quite inexact because the high tide elevation on a particular day can be very different from the mean high tide elevation. On some beaches, there are also easements that authorize public use of the beach landward of the mean high tide line. Members of the public using tidelands should exercise caution when there is a potential for encroaching on private land, just as private land owners should be cautious in questioning the public's right to be on the beach. Persons with questions should contact their local Coastal Commission office or the State Lands Commission in Sacramento.

Finally, the public generally does not have a right to cross private property without permission to get to tidelands. There is an exception, known as the doctrine of implied dedication, or "prescriptive rights." Under this doctrine, the public may acquire the right to cross private property by using the property substantially and continually for at least five years, without permission from the owner, without significant objection by the landowner, and with the landowner's actual or presumed knowledge. The doctrine is a powerful tool, but it does not provide a prompt answer to the casual beachgoer because there must be a legal adjudication to confirm a right of implied dedication.

The Ocean

A WALK ALONG the beach often conjures images of distant travel and far-away voyages. Along much of the California Coast, the view out over the water is of an open expanse of ocean as far as the eye can see, interrupted only by near-shore rocks, ships, or migrating whales. Along the coast of Santa Barbara and Ventura Counties, however, offshore islands are often visible. San Miguel, Santa Rosa, Santa Cruz, and Anacapa Islands, located about 30 miles south of the mainland, and Santa Barbara, San Nicholas, Santa Catalina, and San Clemente Islands farther south make up the Channel Islands. The water between the islands and the mainland is called the Santa Barbara Channel; the islands dramatically influence the coastline and the wave climate throughout this region.

The Pacific Ocean is the largest of the world's oceans; it is larger than all landmasses combined, covering about one-third of the earth's surface. The Pacific is also the deepest of the oceans and, despite the name, it is very stormy. The orientation of the shoreline and the sheltering effect of the islands protect the coast of Santa Barbara and Ventura Counties from most of the storminess. The east-west orientation protects this section of the coast from direct attack by storms moving from the north, and the islands shelter the coast from many of the storms that come up from the south, although coastal areas are still subject to local seas and strong currents running between the islands.

The geologic conditions and circulation patterns in the Santa Barbara Channel support a rich ocean ecosystem. In the late winter and through the spring, nutrient-rich ocean water is pulled to the surface by strong, persistent winds from the south. These cold upwelling areas develop near Point Conception, and the nutrient-rich waters then circulate eastward into the Santa Barbara Channel. In the summer, water on the shallow continental shelf warms and flows westward, into the area of coastal upwelling. The warmer shelf water mixes with the upwelling and often sets up a pattern of counterclockwise circulation, termed convergence. During the summer months, large tongues of cold upwelling water can be detected far to the west of Point Conception, with warmer water being carried south, back into the channel. During the late fall and early winter, the winds shift more toward the east and north, the southern winds relax, and upwelling stops. Warm shelf water continues to flow westward, but when there is no upwelling to set up a recirculation pattern, the warmer shelf water continues flowing to the northwest, around Point Conception, and into the Santa Maria Basin.

While the islands shelter much of the Santa Barbara Channel shoreline from direct wave attack, the shoreline is not protected from all storms. Winter storms associated with Eastern Pacific low-pressure areas create high waves that approach from the west-northwest, attacking the shoreline through gaps between the islands. During El Niño periods, the warmer water temperatures and elevated water levels compound the damaging effect of storm waves. The Central Coast of California received significant damage during El Niño events of 1982–83 and 1997–98. The central California coast also has experienced damaging storms from tropical cyclones—storms that develop off the west coast of Mexico and travel north along the southern California coast. The impacts from these storms tend to be very localized along the coast of the Santa Barbara Channel because the wave energy is selectively dampened or focused as waves refract and propagate through the island gaps. For example, Goleta is most vulnerable to swell from the southwest, whereas the area near the community of Summerland is most vulnerable to waves arriving from the southeast.

The melon-headed **Risso's dolphin** (*Grampus griseus*) is relatively easy to identify at sea. This is especially true of older individuals, due to their distinctive scars and scratches, which deepen with age. These marks are likely caused by the teeth of other Risso's dolphins. Adult Risso's dolphins can reach a length of twelve feet. Risso's dolphins are active at the surface of the water, often seen swimming alongside vessels and surfing the waves. Sometimes they are observed swimming in a line. This formation may improve hunting effectiveness of the group. Risso's dolphins prey primarily on cephalopods such as octopus, cuttlefish, and small squid.

Risso's dolphins

The **harbor porpoise** (*Phocoena phocoena*) is one of the world's smallest cetaceans, growing to a length of about four to six feet. The harbor porpoise is a coastal species, generally inhabiting nearshore waters with a depth of less than 450 feet. The common name of this species is derived from the porpoise's regular appearance in bays and harbors. It is not unusual to see groups of five to ten, lolling about in nearshore waters adjacent to coastal bluffs. Harbor porpoises are not sexually mature until they are between three and five years of age. However, it is estimated that they have a life span of about 20 years. If you get out on the ocean on a whale-watching or fishing trip, it is likely that you will encounter this small and attractive porpoise.

Harbor porpoise

Pacific sardines (*Sardinops sagax*) are small, silvery fishes that grow to a maximum length of 14 inches. Pacific sardines are always seen in schools that may have tens of millions of individuals. Sardines move together as one; this is thought to aid them in predator avoidance. They cover distances of more than 650 miles between feeding and spawning habitats. Pacific sardines are pelagic, meaning that they live in the open ocean. The sardine fishery in California followed a familiar pattern of rapid growth and then accelerating decline. Pacific sardines were harvested with purse seines, supporting the largest fishery in the western hemisphere in the 1930s and 1940s. The fishery collapsed in the 1960s due to heavy exploitation and habitat loss brought on by climate fluctuations. Current population trends indicate that the sardine fishery is recovering.

Pacific sardines

Blue whale

Blue whales (*Balaenoptera musculus*) are the largest animals on Earth. They can grow to 100 feet in length and weigh as much as 150 tons. Lacking teeth, blue whales instead have fringed baleen plates that they use to strain and eat krill, small shrimp-like invertebrates. A blue whale takes a big gulp of water, closes its mouth, pushes the water out through the baleen with its tongue, and then swallows what is left, its catch of krill. Blue whales are known as rorqual whales, meaning those that have pleated grooves that allow their throats to expand during the huge intake of water during filter feeding. Monterey Bay hosts the largest concentration of blue whales in the world. In the summer, whale watchers are often lucky enough to see this majestic giant. Blue whales were hunted nearly to extinction in the early 1900s, but they are currently protected pursuant to the federal Endangered Species Act. It appears that the blue whale population is increasing.

Gray whale

The **gray whale** (*Eschrichtius robustus*) is gray with some blotchy white spots. Numerous barnacles and "whale lice" attach themselves to the surface of the skin. There are few or no parasites on the whale's right side because of the way it scrapes along the ocean bottom to feed. Gray whales feed almost exclusively on benthic, or mud-dwelling, organisms. They sieve through the mud on the bottom of the ocean floor and filter out small crustaceans with their baleen. Gray whales make an extraordinarily long migration each year of 12,000 miles from the feeding grounds in the Bering Sea to the tropical calving lagoons on the Baja California peninsula of Mexico.

Copepod

Copepods may be among the most abundant animals on Earth. Copepods, of which there are many species, are tiny aquatic crustaceans, the diminutive relatives of the crabs and shrimps. They are regarded as the insects of the seas. These microscopic creatures provide a key link in ocean food webs. Planktonic copepods feed on phytoplankton, minute free-floating unicellular algae. In turn, the copepods are eaten by larval fishes and filter feeders, providing the base of the pelagic food chain. *Cope* is the Greek word meaning an "oar" or "paddle"; *pod* is Greek for "foot." This name refers to their broad, paddle-like swimming legs.

The **ocean sunfish** (*Mola mola*) looks a bit like a huge lima bean with its round and flat body. Its body appears to be part of its head, and so it is often called the "headfish" or "moonfish" after its round appearance. It is one of the heaviest bony fish in the world, growing to a length of up to 11 feet and a weight of 5,000 pounds. The ocean sunfish has thick, leathery, scaleless skin protected by a layer of mucous. It has a small mouth, which it uses to slurp in food, such as squid and jellyfish. It shreds the food by spitting it out and sucking it back in until the food is small enough to swallow. If you go out on an ocean tour, look for the unusual ocean sunfish basking at the surface of the water.

Ocean sunfish

California market squid (*Loligo opalescens*) ranges from southeastern Alaska to Baja California. This pelagic mollusk grows to a length of about 12 inches, including its arms and feeding tentacles. It has an elongated body, or mantle, that houses an internal shell, called a pen. The squid is considered an intelligent invertebrate. It also has the ability to change color in an instant to camouflage itself. At one or two years of age, adults migrate to nearshore waters, form dense schools, mate, lay their eggs, and then die. Spawning adults are generally found over sand or mud bottoms at depths of 15 to 130 feet. After the spawning cycle, the squid-fishing season opens. Market squid are vulnerable to large-scale changes in the environment driven by El Niño, and must be managed carefully. California's squid fishery is the largest in the United States.

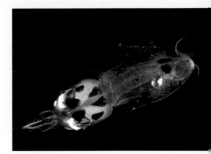

California market squid

The **Pacific loon** (*Gavia pacifica*) is aptly named, for nearly all of the members of this species winter along the Pacific coast. The Pacific loon is a large diving bird that, while floating, rides low in the water. During the breeding season, the Pacific loon makes an eerie yet beautiful call; however, during the winter it is generally quiet. In winter, the diet of the Pacific loon is composed almost entirely of fish. It catches its prey by "surface diving"; the loon floats on the surface searching for underwater prey, and then it quickly dives down in pursuit of a meal. The Pacific loon uses its wings and its legs to propel itself during its underwater pursuits. In addition, the bird has keen eyes that are adapted for underwater vision.

Pacific loon

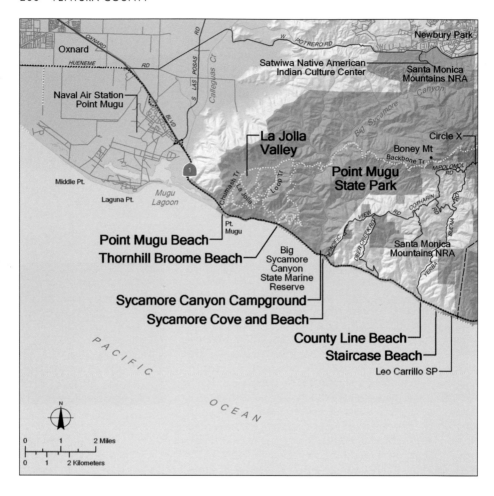

Point Mugu

Point Mugu to Los Angeles County Line	Sandy Beach	Rocky Shore	Trail	Visitor Center	Campground	Wildlife Viewing	Fishing or Boating	Facilities for Disabled	Food and Drink	Restrooms	Parking	Fee
Point Mugu Beach	•	•				•	•		•	•	•	
Point Mugu State Park	•	•	•	•	•	•	•	•	•	•	•	
La Jolla Valley			•			•	•		•	•	•	
Thornhill Broome Beach	•	•	•		•		•	•	•	•	•	
Sycamore Canyon Campground					•			•	•	•	•	
Sycamore Cove and Beach	•			•			•	•	•	•	•	
County Line Beach	•	•				•	•			•	•	
Staircase Beach	•	•	•			•	•			•	•	

POINT MUGU BEACH: *At Point Mugu Rock, off Pacific Coast Hwy.* A sandy beach is located below Pacific Coast Hwy. between Point Mugu Rock and the Navy Firing Range to the north. There is blufftop, off-road parking on the west side of the highway. Portable toilets; lifeguards patrol year-round and staff lifeguard towers in summer. Self-pay stations for day use. There is more parking on the south side of Point Mugu; great views, but no beach access. One-half mile south of Point Mugu is another pull-out, with access to a narrow rocky beach.

POINT MUGU STATE PARK: *Point Mugu E. to Sycamore Canyon.* Point Mugu State Park includes five miles of coastline, inland valleys and peaks, and miles of trails. Big Sycamore Canyon extends seven miles into the Santa Monica Mountains from the shore. Stream courses support riparian woodland vegetation, including western sycamore, big-leaf maple, dogwood, and ferns. Birds in the canyon include red-tailed hawks, woodpeckers, and canyon wrens, and other wildlife includes gray foxes and coyotes. The Department of Parks and Recreation offers wireless Internet access in the park.

LA JOLLA VALLEY: *E. of Hwy. One and Thornhill Broome Beach.* East of Pacific Coast Hwy. in La Jolla Canyon are campsites and a trailhead. Trails offer access to the Boney Mountains State Wilderness Area, located inland on the Backbone Trail that runs east-west through much of the Santa Monica Mountains National Recreation Area. There is trailer or RV camping in the parking lot; picnic tables, tap water, and pit toilets are available. Twelve walk-in campsites with picnic tables, tap water, and chemical toilets are located two miles inland; no fires allowed. Campers should register first at Thornhill Broome Beach campground. Fee for day use and camping.

THORNHILL BROOME BEACH. *Off Pacific Coast Hwy., 1.5 mi. S. of Point Mugu Rock.* Camping only is offered at this unit of Point Mugu State Park, formerly known as La Jolla Beach. There are 68 primitive and unsheltered family campsites located on the beach, adjacent to Pacific Coast Hwy. Chemical toilets; no showers. Bring your own water. Overnight fee. A walk-in group camp for tents only is located on the inland side of Pacific Coast Hwy. All sites can be reserved, year-round; call: 1-800-444-7275. For park information, call: 818-880-0350.

SYCAMORE CANYON CAMPGROUND: *Inland of Pacific Coast Hwy. at Sycamore Cove.* Located in a grove of sycamore trees are 58 developed drive-in campsites with picnic tables and firepits; some sites are wheelchair accessible. Running water; pay showers. An RV dump station is located at the campground. Maximum length for trailers and RVs is 31 feet. Group campsites and hike and bike sites available. Fee for camping. To reserve campsites, call: 1-800-444-7275; for information about Point Mugu State Park, call: 818-880-0350. A trailhead for exploring the Sycamore Canyon and the inland parts of Point Mugu State Park is located here. Except at high tide, there is access to the beach through a pedestrian underpass beneath Pacific Coast Hwy.

SYCAMORE COVE AND BEACH: *Off Pacific Coast Hwy., 4.5 mi. S. of Point Mugu Rock.* A sandy beach 200 yards long is located seaward of Pacific Coast Hwy. at the mouth of Big Sycamore Canyon. Swimming, body surfing, and surf fishing are popular; lifeguards patrol year-round and staff lifeguard towers in summer. There are picnic tables

Sycamore Cove and Beach

The dramatic banded rocks in the vicinity of Point Mugu were deposited in shallow marine waters between 10 and 20 million years ago. Known as the Topanga Formation, the rocks consist mostly of light-gray to tan, moderately hard sandstone, and darker, thinly bedded shale, now uplifted and tilted and well displayed on the southern flank of the Santa Monica Mountains. Containing numerous weak clay layers, these rocks are notoriously unstable, and they host numerous landslides wherever they occur in southern California. The sedimentary rocks are cut by dikes of basalt, part of the Conejo volcanic series that intruded the sedimentary rocks and fed long-eroded volcanoes.

About two and a half miles east of Point Mugu, a huge pile of sand rises to the north of the highway. This sand

has been blown by westerly winds from nearby Thornhill Broome Beach across the highway and against the steep hillside. Drifting sand is a continual maintenance headache for highway crews here and elsewhere along the Pacific Coast Hwy.

Topanga Formation

County Line Beach

and barbecue grills among the trees, paved parking, and restrooms. Call to reserve a beach wheelchair: 805-488-7844. The beach is frequented by western gulls, terns, and migratory shorebirds.

The headquarters for Point Mugu State Park is located here. For information, call: 818-880-0350. There are several signed coastal access points with narrow sandy beaches and blufftop pull-outs to the south along Pacific Coast Hwy., including one at Deer Creek Rd. with stairway beach access. Open 8 AM to sunset; fee for day use.

COUNTY LINE BEACH: *Seaward of Pacific Coast Hwy. at Yerba Buena Rd.* An undeveloped beach area, not actually located at the county line, is located south of Yerba Buena Rd. Unpaved parking area and chemical toilets available. A path leads north along a low bluff overlooking the rocky shore. The beach is a popular spot for surfing, parasailing, and kiteboarding.

STAIRCASE BEACH. *40000 Pacific Coast Hwy., .5 mi. S. of Yerba Buena Rd. and .5 mi. N. of Los Angeles County line.* Look for the street number at the unmarked entrance to the parking lot, which is next door to a state park rangers' residence. Staircase Beach is part of Leo Carrillo State Beach, most of which is in Los Angeles County. A narrow path leads 200 yards down the steep, chaparral-covered bluff to a cobble beach at the base; a broad, sandy beach lies to the south. Chemical toilets are in the parking area.

There are additional beach access points to Leo Carrillo State Beach located two-tenths and one-half mile south of Staircase Beach, along Pacific Coast Hwy. No open fires allowed on the beach. Call: 818-880-0350.

Santa Monica Mountains National Recreation Area

THE SANTA Monica Mountains National Recreation Area is the world's largest urban national park. From Point Mugu in Ventura County, the Recreation Area extends east nearly 50 miles to the Hollywood Bowl, in the heart of metropolitan Los Angeles. The Recreation Area is a cooperative effort, for which planning and management are shared by the National Park Service, the California Department of Parks and Recreation, the Santa Monica Mountains Conservancy, and numerous local governments. Private lands are part of the recreation area, too; the City of Malibu lies entirely within the park's boundary.

The Santa Monica Mountains are part of the Transverse Ranges, running west to east and cut by numerous canyons extending from the high peaks down to the ocean shore. Streams on the canyon floors are bordered by woodlands of oak and sycamore trees. The steep walls of the canyons are characterized by coastal sage scrub and chaparral habitat, largely undisturbed in many places. The Santa Monica Mountains National Recreation Area is notable for this somewhat unusual topography, located within the rare Mediterranean biome. This biome, or ecological community, is characterized by cool, rainy winters and warm, rainless summers, and is found in only five places around the world. The Recreation Area, which lies in close proximity to a huge urban area, is also noteworthy for its biodiversity. Thirteen species of raptors nest in the Santa Monica Mountains, and the park is home to large animals including deer, bobcats, coyotes, and mountain lions. At Malibu Creek, steelhead trout reach the southern limit of their range along the California coast.

Part of the Santa Monica Mountains National Recreation Area is located in Ventura County, with the remaining, larger portion in Los Angeles County. A key component of the Recreation Area in Ventura County is the western end of the system of ridgeline trails known as the Backbone Trail. This trail starts at the Ray Miller Trailhead near Thornhill Broome Beach on Pacific Coast Hwy. in Point Mugu State Park and winds inland about eight miles to the Danielson Multi-use Area, where water and restrooms are

Santa Monica Mountains National Recreation Area

Yerba Buena Road and Boney Mountain

available. The trail continues east about 12 miles to a temporary terminus near the Los Angeles County line. From there, a trail link is planned, but not yet built, to Mulholland Hwy., where the existing Backbone Trail continues east to Castro Crest, through Malibu Creek State Park and Topanga State Park, to Will Rogers State Historic Park.

The Backbone Trail is for day use only; bring plenty of water, and watch out for poison oak and rattlesnakes. Because various segments of the trail have different limitations regarding dogs, trail bikes, and equestrian use, call ahead for rules on the area you plan to visit. The Santa Monica Mountains Trails Council, a nonprofit group working to support the trail system in the mountains, has future plans for trail camps along the Backbone Trail; for information, call: 818-222-4531.

Also in the Ventura County portion of the Santa Monica Mountains National Recreation Area, but reached from Hwy. 101, is the Rancho Sierra Vista, part of a Spanish land grant created in 1803, and the nearby Satwiwa Native American Indian Culture Center, which takes its name from a Chumash Native American village. To reach Rancho Sierra Vista and Satwiwa from Hwy. 101 in Thousand Oaks, take Lynn Rd. south five and a quarter miles to Via Goleta. Dogs must be leashed in this unit of the Santa Monica Mountains National Recreation Area; dogs are not allowed on trails in the adjacent Point Mugu State Park.

Several Ventura County trailheads that provide access to parts of the Recreation Area can be reached from Yerba Buena Rd., which winds inland from Pacific Coast Hwy. just west of County Line Beach. The Canyon View Trail can be reached from Yerba Buena Rd. at a point three-tenths of a mile east of the Circle X group campground and ranger station. No bicycles allowed on the trail. Three-quarters of a mile farther east on Yerba Buena Rd., there is a trailhead for the Sandstone Peak Trail. The top of 3,111-foot Sandstone Peak, reached by a steep three-mile roundtrip hike, offers dramatic vistas of mountains, canyons, and, on a clear day, the Channel Islands. Contact the Santa Monica Mountains National Recreation Area Visitor Center for trail maps and other information. The visitor center is at 401 W. Hillcrest Dr., Thousand Oaks; open 9 AM to 5 PM daily, except federal holidays. Call: 805-370-2301, or see: www.nps.gov/samo.

Protecting Coastal Resources

VISITORS look to the coastline of Monterey, San Luis Obispo, Santa Barbara, and Ventura Counties for beauty, recreation, and inspiration. Visitors come from all around the world, but most are from California, especially from the Los Angeles and San Francisco metropolitan areas, which are like bookends to the Central Coast. This guidebook offers an introduction to some of the myriad recreational learning experiences offered by the coast of central California.

The pressure of growing numbers of residents and visitors brings into focus the finite nature of California's coastal resources. New parks, roads, and other infrastructure may be planned, financed, and built, but the shoreline cannot be expanded. Protecting the coast and restoring lost resources, where possible, are paramount if Californians of the future are to experience the rich array of choices enjoyed by today's coastal visitors.

Each of us who visits the coast can take action to protect and restore the coast's resources (see p. 15). However, many challenges facing the coast require concerted action by society as a whole. None of us alone can successfully address global climate change. As California winters become warmer and wetter in the future, as indicated by current trends, sea level along the shoreline may rise by up to 12 inches. Higher sea levels are likely to cause increased erosion of beaches and cliffs and loss of coastal wetlands, among other changes.

Creation of the California Coastal Trail is an effort that requires the cooperation of many participants. The Coastal Trail is planned as a continuous public right-of-way along the 1,100-mile-long California coastline, from Oregon to Mexico, located as close as possible to the sea. The trail is designed to foster appreciation and stewardship of the scenic and natural resources of the coast through hiking and other complementary modes of non-motorized transportation.

Not just one pathway, the Coastal Trail is more like a yarn composed of several threads, each one a trail alignment or trail improvement that responds to local terrain or accommodates a particular purpose. One thread may be for beach walkers or equestrians,

California Coastal Trail at Pismo Beach, San Luis Obispo County

where a sandy reach of coastline makes this practical. Another segment of the trail may be a paved, urban path for bicyclists, skaters, and wheelchair users. A trail thread may be in use only seasonally, where necessary to skirt a beach on which western snowy plovers nest in spring and summer, or to bypass wintertime high water at a coastal river mouth. An important objective of the Coastal Trail project is to provide a route entirely separate from coastal roads, such as Hwy. One. Because of steep terrain and limited public right-of-way in some locations, however, parts of the trail may require that users share the alignment with vehicles. Each part of the Coastal Trail has its own character, reflecting the great diversity of environments found along California's varied and dynamic shoreline.

About two-thirds of the Coastal Trail has been completed, and work continues. The State Coastal Conservancy, the nonprofit organization Coastwalk, the California Department of Parks and Recreation, and many other groups are working together to make the trail a continuous one and to link parks, ports, communities, schools, trailheads, bus stops, visitor attractions, inns, and campgrounds along the magnificent California coast. For more information on the Coastal Trail including maps for hikers and other trail users, see: www.californiacoastaltrail.info.

Along the Central Coast, a major segment of the California Coastal Trail has been completed. The 29-mile-long Monterey Bay Sanctuary Scenic Trail reaches from Castroville in northern Monterey County to Pacific Grove. Other shorter segments can be found in various locations along the Central Coast. Most anywhere along the shoreline that you are walking on a beach or promenade, or hiking or biking along a paved pathway, you are on the California Coastal Trail.

Preventing habitat loss for plants and creatures is one of the biggest challenges facing California's coast. The consequences of the loss of a species like the majestic bald eagle, symbol of our nation's freedom, are easy to grasp. But lesser known plants and animals also need to be protected from extinction. The biological diversity of living things in an ecosystem is an integral component of that ecosystem's health. The loss of a dominant, or "keystone" species, or the invasion of a nonnative species, can change the way an ecosystem works and challenge the viability of myriad other organisms. Preserving biodiversity is an important management goal for conservation lands, including parks and open space areas.

The Xantus's murrelet (*Synthliboramphus hypoleucus*) is a small bird that spends most of its time at sea along the west coast of North America, from British Columbia to Baja California. Nesting occurs in a much more limited range, on only six of the eight Channel Islands and a few small islands off Baja California. Only one or two eggs are typically placed in a nest, which the Xantus's murrelet makes in a niche in a cliff or in a sea cave. Nonnative rats that had been introduced to the islands took over many of the nest sites formerly used by the birds.

Xantus's murrelet

Studies show that the population of the Xantus's murrelet has declined since the 1970s, indicating the sensitivity of the species to loss of breeding habitat.

The National Park Service, which administers Anacapa Island as part of the Channel Islands National Park, has undertaken efforts to help the Xantus's murrelet. A rat eradication program begun in 2001 has shown small signs of helping the population of Xantus's murrelets to recover. Where the habitat of a species is so limited, as on the Channel Islands, land management decisions can have a dramatic effect on the viability of endangered or threatened plants or animals.

Another Channel Islands animal that has experienced a dramatic decline in population due to habitat loss and predation is the island fox. On Santa Cruz Island, the population of island foxes numbered as many as 2,000 in 1993, while ten years later, only 100 individuals remained. Steps to help the island fox population to recover are being taken, but the program is much more complex than simply removing rats. The island fox's problems include feral pigs and cattle, introduced by 19th century settlers. Foraging by the pigs and livestock grazing have broken down the native coastal sage scrub, chaparral, and oak woodland plant communities, in favor of nonnative annual grasslands, leaving less habitat usable by the island fox.

In the 1990s the foxes, which have no natural predators on the islands, began to be preyed upon by golden eagles, causing the fox populations to plummet. The golden eagles were drawn to the island by the succulent piglets as well as the small foxes, which as adults weigh only three to four pounds. Golden eagles seem to have been able to move into the ecological niche presented by the islands due to the presence of these food sources and also to the absence of the bald eagles that formerly ranged over the islands. Bald eagles are highly territorial, and also typically larger than golden eagles, and the bald eagles tend to dominate their range, to the exclusion of the golden eagles. But bald eagles had vanished from the Channel Islands around 1960, as a result of eating fish contaminated with the pesticide DDT. DDT prevents the eggs of many species of birds, including bald eagles, brown pelicans, and others, from hatching properly. The effect of DDT was devastating to the Channel Island's bald eagles, which fed on fish contaminated by persistent DDT sources in the ocean off southern California. Golden eagles, by contrast, prey on small land mammals and thus escaped the consequences of the pesticide residues.

In a multi-pronged effort, the National Park Service has moved to remove both feral pigs and golden eagles from the four Channel Islands that the service administers. The program has involved some controversy, as both the pigs and the golden eagles have

their adherents, who prefer the status quo, in contrast to Anacapa Island's rats, which had few supporters. A hunting program has reduced the feral pig population. Thirty-two golden eagles have been removed from San Miguel, Santa Rosa, and Santa Cruz Islands and relocated to northern California; none of the relocated birds has returned to the Channel Islands, although others remain. Meanwhile, most of the island foxes have

Island fox

been placed in captive breeding programs on the islands, until their habitat is once again viable for them. The Park Service is also pursuing a program to reintroduce bald eagles to the islands.

In California and throughout the world, biological invasions have become one of the most pressing environmental challenges. Invasive species of plants and animals can reduce native species diversity, change hydrology, and alter the morphology (structure) of habitats. Pampas grass and jubata grass (*Cortaderia* spp.) are invasive coastal plants that grow in disturbed areas, such as cut slopes and bluffs in many locations along the Central Coast, including along Hwy. One in Big Sur. Ironically, pampas grass was popularized in part by a coastal pioneer. Joseph Sexton was a nurseryman who settled in Goleta in the 19th century. As well as growing other plants, Sexton grew pampas grass, which takes little care and is graceful in the garden. Victorian taste favored the plumes, which, in the 1890s, were used to decorate floats in Pasadena's Tournament of Roses Parade.

Like some other invasive plants, pampas grass thrives on disturbed soils, such as in road cuts and clear-cut forest. The plants have very aggressive root systems that overwhelm other plants, even those much larger. Those feathery plumes send their seeds great distances in coastal breezes. Affected habitats include many coastal habitats on the central and southern California coast, including coastal dunes, coastal scrub, Monterey pine forest, and others. The California Department of Parks and Recreation lists pampas grass as one of the top ten most invasive plants in its parks. Efforts have been undertaken to remove pampas grass, such as at Arroyo Burro Beach County Park in Santa Barbara, but the plant seems to be everywhere. One partial solution to the problem of invasive plants would be to require those who create the ground disturbance that enables such species to thrive to take responsibility for the removal and control of the exotic invaders.

Along the streams and rivers of the central and southern California coast, *Arundo donax*, known as giant reed or simply arundo, is another highly invasive plant species. Like

Pampas grass

Arundo

pampas grass, arundo is listed by the California Invasive Plant Council as a most invasive wildland pest plant. Giant reeds grow extremely fast, up to four inches in a day, overrunning and replacing native riparian species, such as sandbar willow. Giant reeds reduce food sources and habitat available to native birds and insects, while providing little of value to native wildlife. The fast-growing plants consume large quantities of water, reducing the supply available to other plants and to wildlife. The tall reeds burn even when green, turning naturally fire-resistant riparian corridors into fire hazards.

Over the past century, there has been a silent invasion of the world's bays and estuaries by alien species, generally carried there in the ballast water of ships, but also arriving with imported clams and oysters or dumped from aquaria by hobbyists who have lost interest. There are over 150 exotic species in San Francisco Bay. Unfortunately, it is not only the large ports and harbors that are suffering this impact. Elkhorn Slough in Monterey County is now home to over 50 nonindigenous colonizing species. One of these, the Japanese mud snail (*Batillaria attramentaria*), was introduced with the planting of Japanese oysters (*Crassostrea gigas*) in the first half of the last century. So far, the mud snail is only found in West Coast estuaries with a history of oyster culturing. Batillaria has thrived at Elkhorn Slough, and now occurs at densities in excess of 5,000 per square meter. The introduction of the mud snail is the likely reason that the native horn snail (*Cerithidea californica*) is now absent from Elkhorn Slough. In other estuaries where the two species currently occur together, the mud snail is slowly displacing the horn snail, a process that takes about 50 years. The mud snail exploits the shared diatom food source more efficiently than does the horn snail.

The continued health and biodiversity of coastal ecosystems depends on the protection and management of high-quality habitat on land and in water, and on an abundance of native species. The health of the coast also depends on sensible land-use decisions that take into account the need for all living creatures to maintain their foothold on the sometimes crowded shores of California. These challenges cannot be met by individuals alone, and neither can they be met by the plant and animal residents of the coast. Instead, awareness and action by all of us are necessary. Next time you go to the coast, enjoy your experience, but learn as much as you can about the places you visit and become aware of what needs to be done to protect these places for future generations to enjoy. As difficult as it

Japanese mud snail, 1.5–2.5 cm long

is to prevent the introduction of exotic species or the loss of habitat or the introduction of poisonous materials into our air and water, it is a simple matter compared to the difficulty of repairing the damage once it has occurred.

Monterey Bay National Marine Sanctuary staff explains tidepool resources to local students

Afterword

In preparing this guide, we visited and researched all beaches, parks, and coastal accessways that are described here. We have incorporated comments and corrections supplied by various agencies and members of the public, who wrote to us regarding the information in previous editions of the *California Coastal Access Guide*. We also sought review of draft material by staff of parks departments, local governments, land trusts, and others. The book is accurate, to the best of our knowledge. Nevertheless, conditions on the coast are constantly changing, and there may be inaccuracies in this book. If you think something is incorrect or has been omitted, please let us know. The Coastal Commission intends to continue publishing revised guides in the future and would appreciate any additional information you can provide. Please remember, however, that this book includes only those beaches and accessways that are managed for public use.

Address all comments to:

Coastal Access Program
California Coastal Commission
45 Fremont Street, Suite 2000
San Francisco, CA 94105

or e-mail to: coast4U@coastal.ca.gov.

Rancho Guadalupe Dunes, Santa Barbara County

Acknowledgments

Additional cartography:

Doug Macmillan

Darryl Rance

Peter Tittmann

Principal Contributors, *California Coastal Access Guide* and *California Coastal Resource Guide*, from which selected material has been incorporated in this book:

Merle Betz

Trevor Kenner Cralle

Linda Goff Evans

Stephen J. Furney-Howe

Jo Ginsberg

Christopher Kroll

Trish Mihalek

Don Neuwirth

S. Briggs Nisbet

Victoria Randlett

Mishell Rose

Sabrina S. Simpson

Pat Stebbins

Mary Travis

Jeffrey D. Zimmerman

Thanks to the following individuals and institutions for their invaluable assistance:

Jack Ainsworth

Joe Barbieri

Will Berg

Shauna Bingham

Robin Blanchfield

Ben Boer

Peter Brand

Jean Bray

Darrell Buxton

California Academy of Sciences

California Resources Agency

Barbara Carey

James Caruso

Kevin Contreras

Susan Craig

Kelly Cuffe

Mark Delaplaine

Jan DiLeo

Tim Duff

Tracy Duffey

Jenifer Dugan

Karren Elsbernd

Phyllis Faber

Claude Garciacelay

Nollie Gildow-Owens

Glenn Greenwald

Rasa Gustaitis

Grey Hayes

Jeff Heiser

Melissa Hetrick

Eileen Hook

Steve Hudson

Rick Hyman

Tim Jensen

Carolyn Johnson

Megan Johnson

James Johnson

Tom Jones

Susan Jordan

Bea Kephart

Tom Killion

Charles Lester

Aileen Loe

Coleen Lund

Sarah D. MacWilliams

Linda McIntyre

Steve Monowitz

Rob Mullane

National Oceanic and Atmospheric Administration

Terri Nevins

Office of Ocean and Coastal Resource Management

Enid Osborn

Lee Otter

Mark Page

John Peirson

Ken Peterson

Deanna Phelps

Alexia Retallack

Willie Richerson

Ed Roberts

Rebecca Roth
Rachel Saunders
Dave Schaechtele
Larry Simon
Gail Skidmore
Greg Smith
Becky Smythe
State Coastal Conservancy
Noah Tilghman
Gary Timm
Misa Ward
Rebecca Young
Mara Ziehn

Photo and Illustration Credits:

© 2002-2007 Kenneth & Gabrielle Adelman, California Coastal Records Project, www.Californiacoastline.org, 192

© AP Images, 258

Rinus Baak, 57

Sherry Ballard © California Academy of Sciences, 201a

Chuck Bancroft, 71b, 161c

Joe Bottomlee, 87a

© Rick Browne, Monterey Bay Aquarium Foundation, 49a

Jeb Byers, 299a

California Historical Society, 45, 66, 100b, 187, 197, 263

California Native Plant Society, 180a

© Mark Conlin/Seapics.com, 128b

Gerald and Buff Corsi © California Academy of Sciences, 71a, 81a, 126a, 137c, 179c

Doug Crapo, 117

E. Craig Cunningham and Malcolm McLeod, 180c

Robert Curtis, 137b

Don DesJardin, 29b, 287c

Joe DiDonato, 88c

Carol DiNolfo, 141

Kay Doyle, 269

Patricia Einstein, 238c

Laura Ann Eliassen, 212a

Kip Evans, front cover, 14, 15, 16, 80a, 80b, 80c, 97, 110, 148c, 200a, 200c, 201b, 201c, 249b, 279b, 286a, 286b

Judith Feins, 234

Dr. Antonio J. Ferreira © California Academy of Sciences, 126c

William Flaxington, 89a

William Follette, 147b, 179b

Grace Fong, 108, 109b

Sonia Garcia, 174

Yohn Gideon, 60

Golden State Aerial Surveys, Inc., 145

Greg Goldsmith, U.S. Fish and Wildlife Service, 147c

Daniel Gotshall, 148a, 250c

Chuck Graham, 178, 243

© Florian Graner/Seapics.com, 285b

James Gratiot, 135a

© 2006 Tom Greer, 88a, 89b, 128c, 160a

Joyce Gross, 29a, 149c, 160b, 160c, 161b, 279c

© David J. Gubernick/www.rainbowspirit.com 26, 109a, 159b

Robert Gustafson, 28a, 135c, 147a, 237c, 249a

Roger Hall, 161a, 239c

© Richard Herrmann/Seapics.com, 62

Richard Herrmann, 250b, 287a

J. N. Hogue, 179a

Beatrice F. Howitt © California Academy of Sciences, 27b, 39c

David Hubbard, 279a

Lloyd Glenn Ingles © California Academy of Sciences, 41a, 127c

Alden Johnson, 71c

Mark Johnsson, 33, 77, 125, 172, 230, 273, 277, 290b

Tom Killion, 10, 78, 106, 198, 228

William Klemens, 76b

Lovell and Libby Langstroth, 287b

Derek Lee, courtesy Hearst Castle/California State Parks, 118a

Sylvie B. Lee, courtesy Hearst Castle/California State Parks, 118b

Ron LeValley, 128a

© Rob Levine, Monterey Bay Aquarium Foundation, 46

David Liebman/Pink Guppy, 40c

Linda Locklin, 294

K. Maupin, Monterey Bay National Marine Sanctuary, 299b

Tom Mikkelsen, 19, 22, 24, 25, 35, 38, 44, 53, 54, 55a, 58, 59, 64, 67, 84, 85a, 90, 95, 114, 120, 122, 138, 140, 142, 153, 158, 164, 166, 168, 176, 177, 186, 193, 194, 196, 202, 204, 231, 232, 235, 274, 276, 280, 282, 288, 290a, 291, 300

Don Millar, 298

© Monterey Bay Aquarium Foundation, 285c, 286c

Keir Morse, 135b

Donald Myrick © California Academy of Sciences, 40b

National Park Service, 295

Oxnard Harbor District (Port of Hueneme), 281

Pacific Gas & Electric Co., 157

Margaret Prokurat, 13

© Todd Pusser/Seapics.com, 63

Dan Richards, National Park Service, 296

Steve Scholl, 27a, 29c, 30, 32, 34, 37, 39b, 42, 43, 47a, 47b, 50, 51, 55b, 56, 69, 70b, 72, 73, 74, 75, 76a, 81b, 82, 85b, 86, 87b, 92, 93, 94, 98, 100a, 104, 105, 111, 115, 121, 124, 127a, 127b, 129a, 129b, 130, 133, 136c, 146, 150, 152, 156, 159c, 162, 165, 169, 170, 181, 205, 206, 207, 208, 209, 210, 211, 213c, 214, 216, 217, 218, 219, 220, 221, 222, 223a, 223b, 224, 237a, 237b, 238a, 253, 256, 259, 260, 262, 264, 268, 270, 278, 285a, 292, 293, 297

Robert V. Schwemmer, West Coast Regional Maritime Heritage Program Coordinator, Channel Islands National Marine Sanctuary, 240, 245, 246, 247, 251, 252, 267

© Andre Seale/Seapics.com, 148b

Dennis Sheridan © David Liebman/Pink Guppy, 213a

Bob Sloan, 238b

Doreen L. Smith, 70c, 87c, 129c, 136a, 136b

Ian Tait, 28b, 81c, 88b, 137a, 149a, 149b, 239a, 239b, 249c

John Tashjian © California Academy of Sciences, 41b

Dean William Taylor, 40a, 134, 236

Ralph Vasquez, U.S. Minerals Management Service, 227

Allan Warner, 151

Charles Webber © California Academy of Sciences, 39a, 159a

© Randy Wilder, Monterey Bay Aquarium Foundation, 48, 250a

Hartmut Wisch, 180b

Carol Witham, 212b, 213b

© David Wrobel/Seapics.com, 200b

© Rick York, 70a, 126b

Glossary

alluvium. Stream deposits and sediments formed by the action of running water.

anadromous. Migrating from salt water to fresh water in order to reproduce.

annual. A plant that germinates, flowers, sets seed, and dies within one year or less.

basalt. A dark igneous rock of volcanic origin. Basalt is the bedrock of most of the world ocean.

bay. A partially enclosed inlet of the ocean.

beach. The shore of a body of water, usually covered by sand or pebbles.

biota. The collective plant and animal life, or flora and fauna, of a region.

bivalves. Mollusks such as clams and oysters that have two-piece, hinged shells.

bluff. A high bank or bold headland with a broad, precipitous, sometimes rounded cliff face overlooking a plain or a body of water.

brackish water. Water that contains some salt, but less than sea water (from 0.5 to 30 parts per thousand).

bunchgrass. Perennial grass that forms tufted clumps; includes many native California species, such as needlegrass, bentgrass, reedgrass, and hairgrass.

cetaceans. A group of aquatic mammals including whales, dolphins, and porpoises.

chert. A fine-grained siliceous sedimentary rock; a source of flint used by California Indians to make spear and arrow blades.

closed-cone. Refers to coniferous trees having cones that remain closed for several or many years after maturing; e.g., bishop pine.

coastal scrub. A plant association characterized by low, drought-resistant, woody shrubs; includes coastal sage scrub.

coastal strand community. A plant association endemic to bluffs, dunes, and sandy beaches and adapted to saline conditions; includes sea rocket and sand verbena.

coastal terrace. A flat plain edging the ocean; uplifted sea floor that was cut and eroded by wave action. Also called marine terrace or wave-cut bench.

conifer. A cone-bearing tree of the pine family, usually evergreen.

continental shelf. The shallow, gradually sloping area of the sea floor adjacent to the shoreline, terminating seaward at the continental slope.

crustaceans. A group of mostly marine arthropods, e.g., barnacles, shrimp, and crabs.

cryptic. Tending to conceal or camouflage.

current. Local or large-scale water movement that results in the flow of water in a particular direction, e.g., alongshore, or offshore.

delta. A fan-shaped alluvial deposit at the mouth of a river.

diapause. A period during which growth or development of an organism is suspended.

El Niño. A warming of the ocean current along the coasts of Peru and Ecuador that is generally associated with dramatic changes in the weather patterns around the world.

endangered. Refers specifically to those species designated by the California Dept. of Fish and Game or the U.S. Fish and Wildlife Service as "endangered" because of severe population declines.

endemic. A plant or animal native to a well-defined geographic area and restricted to that area.

erosion. The gradual breakdown of land by weathering, solution, corrosion, abrasion, or transportation, caused by action of wind, water, or ice; the opposite of accretion.

estero. Spanish for estuary, inlet, or marsh.

estivate. To pass the summer in a state of dormancy.

estuary. A semi-enclosed coastal body of water that is connected with the open ocean and within which sea water mixes with fresh water from a river or stream.

exotic. Any species, especially a plant, not

native to the area where it occurs; introduced.

fault. A fracture or fracture zone along which displacement of the earth occurs resulting from tectonic activity.

foredune. Dune closest to the seashore that is relatively unstable and subject to salt spray, wind erosion, and storm waves; generally sparsely vegetated by plants with special adaptations to harsh conditions.

Franciscan Complex. A group of sedimentary and volcanic rocks that occurs along much of the Northern and Central California Coast and that consists predominantly of sandstone, shale, and chert, with occasional limestone, basalt, serpentine, and schist, formed in the Jurassic and Cretaceous periods.

gneiss. A coarse-grained, banded metamorphic rock, usually of the same composition as granite, in which the minerals show a strong orientation.

groin. A low, narrow jetty, constructed at right angles to the shoreline, that projects out into the water to trap sand or to retard shoreline erosion; a shoreline protective device.

gyre. A large circular or spiral motion of currents around an ocean basin.

habitat. The sum total of all the living and nonliving factors that surround and potentially influence an organism; a particular organism's environment.

halophyte. A plant that is adapted to grow in salty soils.

haul-out. A place where pinnipeds emerge from the water onto land to rest or breed.

intertidal. The shoreline area between the highest high tide mark and the lowest low tide mark.

intrusion. In geology, the process of emplacement of molten rock in pre-existing rock.

invasive species. Weedy plants or wildlife species that colonize and proliferate in areas where they do not naturally occur.

invertebrate. An animal with no backbone or spinal column; 95 percent of the species in the animal kingdom are invertebrates.

jetty. An engineered structure constructed at right angles to the coast at the mouth of a river or harbor to help stabilize the entrance; usually constructed in pairs on each side of a channel.

La Niña. A periodic cooling of surface ocean waters in the eastern tropical Pacific along with a shift in convection in the western Pacific, affecting weather patterns around the world.

lagoon. A body of fresh or brackish water separated from the sea by a sandbar or reef.

longshore current. A current flowing parallel to and near shore that is the result of waves hitting the beach at an oblique angle.

marsh. General term for a semi-aquatic area with relatively still, shallow water, such as the shore of a pond, lake, or protected bay or estuary, and characterized by mineral soils that support herbaceous vegetation.

mollusks. Soft-bodied, generally shelled invertebrates; for example, chitons, snails, limpets, bivalves, and squid.

nearshore. The area extending seaward from the shoreline through the breaker zone to the area where deepwater waves begin to be influenced by the sea floor.

pelagic. Pertaining to open ocean rather than inland waters or waters adjacent to land.

perennial. A plant that lives longer than a year.

pinnipeds. Marine mammals that have fin-like flippers, including seals, sea lions, and walruses.

plankton. Free-floating algae (phytoplankton) or animals (zooplankton) that drift in the water, ranging from microscopic organisms to larger species such as jellyfish.

predator. An animal that eats other animals; a carnivore.

raptors. Birds of prey, such as falcons, eagles, and owls.

reef. A submerged ridge of rock or coral near the surface of the water.

relict. In ecology, a genus or species from a previous era that has survived radical environmental changes resulting from climatic shifts.

revetment. A sloped retaining wall built of rip-rap or concrete blocks along the coast to prevent erosion inland and other damage by wave action; similar to a seawall.

rip current. A narrow, swift-flowing current that flows seaward through the breaker zone at nearly right angles to the shoreline and returns water to the sea after being piled up on the shore by waves and wind.

riparian. Pertaining to the habitat along the bank of a stream, river, pond, or lake.

riprap. Boulders or quarry stone used to construct a groin, jetty, or revetment.

rookery. A breeding site, such as an island, for seabirds or marine mammals.

schist. Medium- to coarse-grained metamorphic rocks composed of laminated, often flaky, parallel layers of chiefly micaceous minerals.

sea stack. A tall island of rock left standing after waves have eroded the shoreline.

seawall. A structure, usually a vertical wood or concrete wall, designed to prevent erosion inland or damage due to wave action.

sedimentary rock. Rocks resulting from the consolidation of loose sediment.

serpentine. A green or black magnesium-silicate mineral, or a rock composed principally of serpentine. Nutrient-poor soils of serpentine origin support a unique association of endemic plants.

shale. A fine-grained sedimentary rock formed by the consolidation of clay, silt, or mud.

siliceous. Containing or consisting of silica.

slough. A small marshland or tidal waterway that usually connects with other tidal areas.

species. A taxonomic classification ranking below a genus, and consisting of a group of closely related organisms that are capable of interbreeding and producing viable offspring.

substrate. The surface on which an organism grows or is attached.

surf zone. The area affected by wave action, from the shoreline high-water mark seaward to where the waves start to break.

take. As defined by the Endangered Species Act, "to harass, harm, pursue, hunt, shoot, wound, kill, capture, or collect, or attempt to engage in any such conduct."

terrestrial. Living or growing on land, as opposed to living in water or air.

threatened. Refers specifically to those species designated by the California Dept. of Fish and Game or the U.S. Fish and Wildlife Service as "threatened" because of severe population declines.

tidal wave. The regular rise and fall of the tides; often misused for tsunami.

tide. The periodic rising and falling of the ocean resulting from the gravitational forces of the moon and sun acting upon the rotating earth.

tidepool. Habitat in the rocky intertidal zone that retains some water at low tide.

tombolo. A bar of gravel or sand connecting an island or rock with the mainland or with another island.

tsunami. A sometimes destructive ocean wave caused by an underwater earthquake, submarine landslide, or volcanic eruption; inaccurately called a tidal wave.

understory. A layer or level of vegetation occurring under a vegetative canopy, such as the herbaceous plants growing under the taller trees of a riparian woodland.

uplifted. Pertaining to a segment of the earth's surface that has been elevated relative to the surrounding surface as a result of tectonic activity.

upwelling. A process by which deep, cold, nutrient-rich waters rise to the sea surface.

waterfowl. Ducks, geese, and swans.

watershed. The land area drained by a river or stream system or other body of water.

wetland. General term referring to shallow water (less than six feet deep) and land that is tidally or seasonally inundated, including marshes, mudflats, lagoons, sloughs, bogs, swamps, and fens.

Selected State and Federal Agencies

California State Agencies:

California Coastal Commission
725 Front St., Suite 300
Santa Cruz, CA 95060
831-427-4863

California Coastal Commission
89 South California St., Suite 200
Ventura, CA 93001
805-585-1800

California Department of Fish and Game
1416 Ninth St.
Sacramento, CA 95814
916-445-0411

California Department of Fish and Game
Marine Region Field Office
20 Lower Ragsdale Dr., Suite 100
Monterey, CA 93940
831-649-2870

California Department of Fish and Game
Marine Region Field Office
213 Beach St.
Morro Bay, CA 93442
805-772-3011

California Department of Fish and Game
Marine Region Field Office
1933 Cliff Dr., Suite 9
Santa Barbara, CA 93109
805-568-1231

California Department of Parks and Recreation
1416 Ninth St.
Sacramento, CA 95814
1-800-777-0369

State Coastal Conservancy
1330 Broadway, Suite 1100
Oakland, CA 94612
510-286-1015

California State Lands Commission
100 Howe Ave., Suite 100 South
Sacramento, CA 95825-8202
916-574-1900

Federal Agencies:

Channel Islands National Marine Sanctuary
113 Harbor Way, Suite 150
Santa Barbara, CA 93109
805-966-7107

Channel Islands National Marine Sanctuary
3600 South Harbor Blvd., Suite 111
Oxnard, CA 93035
805-382-6151

Channel Islands National Park
1901 Spinnaker Dr.
Ventura, CA 93001
805-658-5730

Los Padres National Forest
6755 Hollister Ave., Suite 150
Goleta, CA 93117
805-968-6640

Monterey Bay National Marine Sanctuary
299 Foam St.
Monterey, CA 93940
831-647-4202

Santa Monica Mountains National
Recreation Area
401 West Hillcrest Dr.
Thousand Oaks, CA 91360
805-370-2301

U.S. Fish and Wildlife Service
2493 Portola Rd., Suite B
Ventura, CA 93003
805-644-1766

Bibliography

Albert, Ken. *Fishing in Southern California: The Complete Guide.* Huntington Beach, CA: Marketscope, 2003.

Peter Browning (editor). *The Discovery of San Francisco Bay: The Portola Expedition of 1769-1770 - The Diary of Miguel Costansó.* Lafayette, CA: Great West Books, 1992.

California Coastal Commission. *California Coastal Access Guide.* 6th ed. Berkeley: University of California Press, 2003.

————. *California Coastal Resource Guide.* Berkeley: University of California Press, 1987.

Cooper, Daniel S. *Important Bird Areas of California.* Pasadena: Audubon California, 2004.

Dana, Richard Henry, with introduction by Gary Kinder and notes by Duncan Hasell. *Two Years before the Mast: A Personal Narrative of Life at Sea.* New York: Modern Library, 2001.

Delgado, James P. *To California by Sea: A Maritime History of the California Gold Rush.* Columbia: University of South Carolina Press, 1990.

Felton, Ernest L. *California's Many Climates.* Palo Alto, CA: Pacific Books, 1965.

Goodson, Gar. *Fishes of the Pacific Coast.* Stanford, CA: Stanford University Press, 1988.

Griggs, Gary, Kiki Patsch, and Lauret Savoy. *Living with the Changing California Coast.* Berkeley: University of California Press, 2005.

Gudde, Erwin G. *California Place Names: The Origin and Etymology of Current Geographical Names.* Berkeley: University of California Press, 2004.

Henson, Paul, and Donald. J. Usner. *The Natural History of Big Sur.* Berkeley: University of California Press, 1993.

Hoover, Mildred Brooke, Hero Eugene Rensch, and Ethel Grace Rensch. *Historic Spots in California.* 3rd ed. Stanford, CA: Stanford University Press, 1966.

Kampion, Drew, ed. *The Stormrider Guide: North America.* Bude, Cornwall, UK: Low Pressure Ltd., 2002.

Lanner, Ronald M. *Conifers of California.* Los Olivos, CA: Cachuma Press,1999.

Lynch, David K. and William Livingston. *Color and Light in Nature.* Cambridge: Cambridge University Press, 1995.

Manesson-Mallet, Allain. *Description de l'univers* (Tome V), *Nouveau Mexique et Californie* (map). Paris, 1683.

Marinacci, Barbara and Rudy Marinacci. *California's Spanish Place Names: What They Are and How They Got Here.* San Rafael, CA: Presidio Press, 1980.

McLaughlin, Glen, with Nancy H. Mayo. *The Mapping of California as an Island: An Illustrated Checklist.* Saratoga, CA: California Map Society, 1995.

Morris, William, ed. *The American Heritage Dictionary of the English Language.* Boston, MA: Houghton Mifflin Co., 1976.

Munz, Philip A. *Introduction to Shore Wildflowers of California, Oregon, and Washington.* Berkeley: University of California Press, 2003.

Pavlik, Bruce M., Pamela C. Muick, Sharon G. Johnson, and Marjorie Popper. *Oaks of California.* 5th printing with revisions. Los Olivos, CA: Cachuma Press and the California Oak Foundation, 2002.

Signor, John R. *Southern Pacific's Coast Line.* Wilton, CA: Signature Press, 1994.

Stuart, John D., and John O. Sawyer. *Trees and Shrubs of California.* Berkeley: University of California Press, 2001.

Tompkins, Walker A. *Santa Barbara Yesterdays.* Facsimile reprint. Santa Barbara, CA: McNally and Loftin Publishers, 1996.

Uhrowczik, Peter. *The Burning of Monterey: The 1818 Attack on California by the Privateer Bouchard.* Los Gatos, CA: Cyril Books, 2001.

Wheeler, Eugene D., and Robert E. Kallman. *Shipwrecks, Smugglers and Maritime Mysteries.* 3rd ed. Ventura, CA: Pathfinder Publishing, 1989.

Suggestions for Further Reading:

Bakker, Elna S. *An Island Called California: An Ecological Introduction to Its Natural Communities.* 2nd ed. Berkeley: University of California Press, 1984.

Bascom, Willard. *Waves and Beaches: The Dynamics of the Ocean Surface.* Garden City, NY: Anchor Press, 1980.

California Coast & Ocean, a quarterly magazine covering trends, issues, and controversies shaping the California coast; see www.coastandocean.org.

Cralle, Trevor. *Surfin'ary: A Dictionary of Surfing Terms and Surfspeak.* Rev. ed. Berkeley, CA: Ten Speed Press, 2001. Surfers' lingo and technical terms, augmented by history of the sport.

Garth, John S., and J. W. Tilden. *California Butterflies.* Berkeley: University of California Press, 1986.

Guisado, Raul, and Jeff Klaas. *Surfing California: A Complete Guide to the Best Breaks on the California Coast.* Guilford, CT: Falcon Guide, 2005.

Humann, Paul. *Coastal Fish Identification: California to Alaska.* Jacksonville, FL: New World Publications, 1996.

Jones, Ken. *Pier Fishing in California: The Complete Coast and Bay Guide.* 2nd ed. Roseville, CA: Publishers Design Group, 2004.

Langstroth, Lovell, and Libby Langstroth. *A Living Bay: the Underwater World of Monterey Bay.* Berkeley: University of California Press, 2000. Monterey Bay Aquarium Series in Marine Conservation.

Love, Milton. *Probably More Than You Want to Know about the Fishes of the Pacific Coast.* Santa Barbara, CA: Really Big Press, 1996.

McPeak, Ronald H., Dale A. Glantz, and Carole Shaw. *The Amber Forest: Beauty and Biology of California's Submarine Forests.* San Diego, CA: Watersport Publications, 1988.

Mondragon, Jennifer, and Jeff Mondragon. *Seaweeds of the Pacific Coast: Common Marine Algae from Alaska to Baja California*. Monterey, CA: Sea Challengers, 2003.

Monterey Bay Aquarium, with the National Oceanic and Atmospheric Administration. *A Natural History of the Monterey Bay National Marine Sanctuary*. Monterey, CA: Monterey Bay Aquarium Foundation, 1997.

Powell, David C. *A Fascination for Fish: Adventures of an Underwater Pioneer*. Berkeley: University of California Press, 2001. Monterey Bay Aquarium Series in Marine Conservation.

Ricketts, Edward F., Jack Calvin, and Joel Hedgpeth. *Between Pacific Tides*. Rev. ed. Stanford, CA: Stanford University Press, 1992. For generations, a classic in the field.

Schoenherr, Allan A. *A Natural History of California*. Berkeley: University of California Press, 1992.

Schoenherr, Allan A., C. Robert Feldmeth, and Michael J. Emerson. *Natural History of the Islands of California*. Berkeley: University of California Press, 1999.

Starr, Kevin. *California: A History*. New York: Modern Library, 2005.

Stallcup, Rich. *Ocean Birds of the Nearshore Pacific: A Guide for the Sea-Going Naturalist*. Stinson Beach, CA: Point Reyes Bird Observatory, 1990.

Williams, Glyn. *Voyages of Delusion: The Quest for the Northwest Passage*. New Haven, CT: Yale University Press, 2003.

For Younger Readers:

Aykroyd, Clarissa. *Exploration of the California Coast: The Adventures of Juan Rodríguez Cabrillo, Francis Drake, Sebastián Vizcaíno, and Other Explorers of North America's West Coast*. Philadelphia, PA: Mason Crest Press, 2003.

Kalman, Bobbie. *Life of the California Coast Nations*. Crabtree Publishing Company, 2004. Day-to-day life along the pre-Columbian California coast.

León, Vicki. Photography by Richard Bucich and Jeff Foott. *A Raft of Sea Otters*. Parsippany, NJ: Silver Burdett Press, 1995.

St. Antoine, Sara. *The California Coast: Stories from Where We Live*. Minneapolis: Milkweed Editions, 2001. A collection of essays, letters, poems, and stories that celebrate and explore life on the California Coast.

Sobol, Richard. *Adelina's Whales*. New York: Dutton Children's Books, 2003.

Steinbeck, John. *Cannery Row*. John Steinbeck Centennial Edition. New York: Penguin Books, 2002.

Index